国家自然科学基金面上项目（项目批准号：51378359）

Urban Planning and Design Course Facing International Students

面向国际学生的城市规划与设计教程

田 莉 李 晴 等著
Li Tian and Qing Li ed.

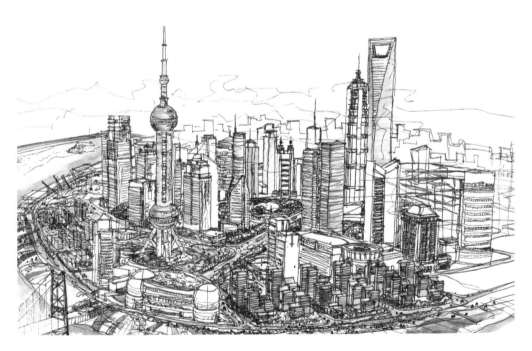

其他著者：Richard Dagenhart, Stefan Al, 陈竞姝
Other contributors: Richard Dagenhart, Stefan Al, and Jingshu Chen

中国建筑工业出版社
CHINA ARCHITECTURE & BUILDING PRESS

图书在版编目（CIP）数据

面向国际学生的城市规划与设计教程 / 田莉等著. —北京：中国建筑工业出版社，2015.12
ISBN 978-7-112-19001-0

Ⅰ.①面… Ⅱ.①田… Ⅲ.①城市规划–建筑设计–教材 Ⅳ.①TU984

中国版本图书馆CIP数据核字（2016）第010389号

责任编辑：杨　虹
责任校对：陈晶晶　李美娜

面向国际学生的城市规划与设计教程
田　莉　李　晴　等著
*
中国建筑工业出版社出版、发行（北京海淀三里河路9号）
各地新华书店、建筑书店经销
北京嘉泰利德公司制版
北京缤索印刷有限公司印刷
*
开本：787×1092毫米　1/16　印张：11　字数：350千字
2017年1月第一版　2017年1月第一次印刷
定价：58.00元
ISBN 978-7-112-19001-0
　　　（28283）

版权所有　翻印必究
如有印装质量问题，可寄本社退换
（邮政编码 100037）

近年来，随着我国在学术界国际交流的日益繁荣，越来越多的国际学生到中国大学里攻读学位或联培双学位，但目前我国尚未有一本可以满足国际学生需求的英文城市设计教材。经过多年来中外高校联合城市设计的实践及2010年以来开设面对国际学生的城市规划与设计课程的经验，同济大学建筑城规学院的田莉教授和李晴副教授对课程建设和授课方法进行了总结，并邀请海外知名大学教授参与，编写了该本中英文双语的城市设计教材。内容包括基地研究与分析方法，概念生成和成果要求等，并选择了四期有代表性的学生设计成果，以期为越来越多的城市规划院校开设的面向国际学生的城市规划与设计课程提供教学参考，使不同背景下的国际学生可以较快地掌握我国快速城市化背景下的城市设计基础方法，了解课程的需求，并适度了解和参与中国的设计实践。

With the burgeoning international communication in the academic circle of China, more and more overseas students come to China to pursue degrees of planning and design in Chinese universities or join the dual-degree programs of Chinese and overseas universities. However, a bilingual English-Chinese textbook of planning and design which can meet the needs of international students has been absent in China, and this book is such an attempt to fill this gap. After several years of teaching joint-design programs and design courses for international students, Prof. Li Tian and Qing Li from Tongji University compile this book to introduce the curriculum and teaching methods of planning and design course for international students. Meanwhile, they invite overseas well-known professors from top universities in the world to share their teaching experience. The content of this book includes site survey and analytical methods, design concept development and requirements for final production. Then it selects the student works of four design projects as references for other students. In general, this book aims to help international students know the urban planning and design approach under the backdrop of rapid urbanization of China, and provide them chances to understand and participate in local planning and design practice.

Contents

Preface

Part I : Introduction to Course/1

A Brief Introduction of the Urban Planning and Design Course for International Students/2
- 1.1 A Brief Introduction of the English Urban Planning and Design Course/2
- 1.2 The Curriculum Framework of the English Urban Planning and Design Course/2
- 1.3 Teaching Approach of the English Urban Planning and Design Course/4
- 1.4 Conclusion/6

Part II : Teaching Methods、Process and Final Production/9

- 2.1 Urban Planning & Design for International Students: Site Study and Approaches/10
  - 2.1.1 Levels and Contents of a Site Study/10
  - 2.1.2 Common Approaches of a Site Study/12
  - 2.1.3 Site Study Results/14

- 2.2 Generation Framework of Urban Planning and Design Concepts/18
  - 2.2.1 Preliminary Study/18
  - 2.2.2 Field Research/18
  - 2.2.3 Eidetic Universal/18
  - 2.2.4 Preliminary Concept/20
  - 2.2.5 Conceptual Analysis/20

- 2.3 Urban Planning and Design for International Students: Production Requirements for Design Outcomes/24
  - 2.3.1 Goal of Design Production/24
  - 2.3.2 The Content of Design Production/24
  - 2.3.3 Expression Modes of Design Production/28
  - 2.3.4 Review of Design Production/30

Part III : International Experiences of Urban Design Studio Teaching/33

- 3.1 Urban Design Pedagogies in an Increasingly Globalized World/34
- 3.2 Urban Design Studio at Georgia Tech: University Ave, Pittsburgh Neighborhood and the Mcdaniel Creek Watershed/38

Part IV : Collections of Works of International Students/45

- 4.1　Urban Design of Lujiazui Central Business District Regeneration/46
  - 4.1.1　Lujiazui : Real Estate Game/48
  - 4.1.2　Endless City : Ecocity /54
  - 4.1.3　City of Mirror/60
  - 4.1.4　The Public Domain & The City of Movement/66
  - 4.1.5　Playful City Playful Life/72

- 4.2　Conceptual Urban Design of Coastal Areas in Jinshan District, Shanghai/78
  - 4.2.1　Space Metamorphosis : FLOATING ISLAND/80
  - 4.2.2　Organic Urbanism/86
  - 4.2.3　Back to the Sea/90
  - 4.2.4　Living From the Wall to Rainbow/98
  - 4.2.5　Level of Games/102

- 4.3　Urban Design of Central Area of Tinglin Township, Jinshan District, Shanghai/108
  - 4.3.1　Water Side Life/110
  - 4.3.2　Metamorphosis/116
  - 4.3.3　That's Tinglin/122
  - 4.3.4　Tingreen/132

- 4.4　Urban Design of International Cruise Terminal Area, Baoshan District, Shanghai/140
  - 4.4.1　Baoshan: The Prism City/142
  - 4.4.2　Along The River/148
  - 4.4.3　Boat Shan/154
  - 4.4.4　Baoshan Waterland City/160

目录

前言

第一部分：课程简介 /1

面向国际留学生的城市规划与设计全英语课程建设简介 /2
    1.1 全英文城市规划与设计课程建设概况 /3
    1.2 全英文城市规划与设计课程建设框架 /3
    1.3 全英文城市规划与设计课程教学方法 /5
    1.4 总结 /7

第二部分：教学方法、过程和成果要求 /9

2.1   面向国际留学生的城市设计：基地分析与研究方法 /10
    2.1.1 基地分析的层次和内容 /11
    2.1.2 基地分析与研究的常见方法 /13
    2.1.3 基地分析的成果构成 /15

2.2   城市规划设计概念生成框架 /18
    2.2.1 初步研究 /19
    2.2.2 现场调研 /19
    2.2.3 本质共相 /19
    2.2.4 初步概念 /21
    2.2.5 概念解析 /21

2.3   面向国际留学生的城市规划与设计：设计成果制作要求 /24
    2.3.1 设计成果的目标 /25
    2.3.2 设计成果的内容 /25
    2.3.3 设计成果的表达方式 /29
    2.3.4 设计成果的评审 /31

第三部分：城市设计教学的国际经验 /33

    3.1 在日益全球化的世界中的城市设计教学法 /34
    3.2 佐治亚理工大学城市设计教学案例：匹兹堡社区的大学路和麦克丹尼尔河流域设计 /38

第四部分：国际学生设计作业选编 /45

4.1　上海陆家嘴中心区更新概念城市设计 /46
- 4.1.1　陆家嘴——房地产的游戏 /48
- 4.1.2　永续的城市——生态城市 /54
- 4.1.3　镜像城市 /60
- 4.1.4　公共空间与流动城市 /66
- 4.1.5　乐活城市，乐活生活 /72

4.2　上海金山区城市生活岸线概念性城市设计 /78
- 4.2.1　空间的变异——浮游的岛屿 /80
- 4.2.2　有机都市主义 /86
- 4.2.3　回归海洋 /90
- 4.2.4　彩虹带 /98
- 4.2.5　立体游戏 /102

4.3　上海金山区亭林镇中心区城市设计 /108
- 4.3.1　水岸生活 /110
- 4.3.2　蜕变 /116
- 4.3.3　很亭林 /122
- 4.3.4　植绿亭林 /132

4.4　上海宝山区国际邮轮港地区城市设计 /140
- 4.4.1　宝山：棱镜城市 /142
- 4.4.2　河畔之城 /148
- 4.4.3　船"山" /154
- 4.4.4　宝山水陆都市 /160

Preface

We are in an era of globalization. But local characteristics remain one of the most emphasized parts in today's urban planning and design. When well-developed countries entered into the stage of matured urban development, the physical space available to urban planners there is growing increasingly narrow. On the other hand, since the implementation of the reform and opening policy, China began experiencing a period of fast urbanization, providing urban planners and designers all over the world and at home with a vast new arena, which has attracted thousands of international students and designers to China. However, how to develop an urban planning curriculum that combines local context and global view poses a great challenge not only to international students and designers who are unfamiliar with the unique political, economic and social environment and the government-led development mode of China, but to Chinese educators as well. Based on the practices we've developed when cooperating with foreign universities on many joint urban planning projects and, especially, our experience in creating urban planning and design courses for international students since 2010, the College of Architecture and Urban Planning(CAUP), Tongji University, carefully refined our curriculum structure and teaching approaches and compiled a new Chinese-English bilingual text book. By presenting this new text book, we hope it could provide a good reference for domestic urban planning colleges that are intended to open similar courses for international students and, especially, as more educators and students moving along this carefully designed course, we could receive more feedbacks and accumulate more experiences and materials for further improvement of it.

In addition to our own teachers, we invited designers from Shanghai Tongji Urban Planning & Design Institute into our teaching team, helping students broaden their horizon and give them practical instruction during their design practice. Also, we invited some prominent professors from universities with global influence and prestige in this field, for example, Prof. Richard Dagenhart, from City & Regional Planning School, College of Architecture, Georgia Institute of Technology, USA, who led the Tongji-Georgia Tech Joint Urban Design Project for several years, Prof. Stefan Al from Department of City Planning, School of Design, University of Pennsylvania, USA, who used to teach city planning courses in the Department of Urban Planning and Design, the University of Hong Kong, HK SAR, to write some chapters for this Text book and bring us international teaching experience. Prof. Zheng Zheng, former director of urban design program, Urban Planning Department, Tongji University, also made a great contribution to this book with his rich experience in teaching and design field.

The text book is made up of following four parts:

Part I is a brief introduction of the English Urban Planning and Design Course for international students, including the structure, content and teaching approaches of the course.

Part II consists of three components, i.e. Site Analysis and Methodologies, Concept Deveopment, as well as the Requirements on the Production of Final Result.

Part III is an introduction made by overseas professors about their teaching experiences and practices in the city planning.

Part IV is a fine collection of excellent urban design works of international students since 2010.

We wish to express our deep gratitude to Ms. Jingwei Li, Mr. Kun Du, Mr. Rui Yang and Mr. Bo Zhang, postgraduates of Urban Planning Department, Architecture and Urban Planning School, Tongji University, and Ms. Jie Zhou from Shanghai Tongji Urban Planning & Design Institute for their kindly help in typesetting and material collection and sorting. We are deeply indebted to Ms. Hong Yang, Editor from China Architecture & Building Press. It is impossible for us to publish text book without her valuable supports and kind help. The writers take responsibility for their views.

Li Tian, Qing Li, Oct. 2015

前 言

我们正处于一个全球化的时代，然而规划设计带有强烈的地方色彩。在发达国家进入成熟的城市化阶段，物质性规划设计空间日益压缩的时候，改革开放以来处于快速城镇化进程的中国，已成为国际设计师和本国设计师的"竞技场"，并吸引了数以万计的海外留学生来此学习和工作。然而，在中国"政府主导"的发展模式及独特的政治经济社会发展环境下，海外留学生在短短一个学期的时间内，面对不熟悉的本地社会、经济及文化背景，如何能构思出切合本地语境、又具有国际视野的城市设计方案，是任课教师和学生所面临的挑战。经过多年来中外高校联合城市设计的实践及2010年以来开设面对国际留学生的城市规划与设计课程的经验，同济大学建筑城规学院田莉教授和李晴副教授对课程建设和授课方法进行了总结，编写了本中英文双语的教材，抛砖引玉，以期为越来越多的城市规划院校开设的面向国际学生的城市规划与设计课程提供参考，并希冀随着设计教学的进一步开展，积累更多的经验，为将来完善本教材提供更多的素材。

除了参与任课的教师外，我们还邀请了上海同济城市规划设计研究院的设计师参与教学工作，为拓展学生视野、贴近地方实践提供帮助。另外，我们邀请了国际知名院校的知名教授，包括美国佐治亚理工大学建筑与规划学院负责城市设计方向的 Richard Dagenhart 教授（他多年负责同济－佐治亚理工的联合城市设计）、宾州大学城市规划系的 Stefan Al 教授（他之前在香港大学城市规划系具有多年的城市设计教学经验）为本书撰写相关的章节，提供具有国际视野的教学经验。同济大学前城市规划系城市设计教研室主任、具有丰富教学和实践经验的郑正教授为本书的撰写提供了宝贵的意见。

本书的构成包括如下四个板块：

第一板块为面向国际留学生的全英语城市规划与设计课程建设简介，系统介绍课程建设的框架、内容和教学方法。

第二板块分为三部分，分别包括基地分析与研究方法、概念生成和最终成果制作要求。

第三板块为国外教授介绍其城市设计的教学经验和实践。

第四板块为2010年以来国际留学生城市设计作业的成果精选。

本书的排版和资料整理工作由同济大学建筑城规学院城市规划系研究生李经纬、杜坤、杨瑞、张博和上海同济城市规划设计院邹洁承担，中国建工出版社的杨虹编辑为本书的出版提供了大力支持和帮助，没有他们本书的出版是难以想象的，在此一并致谢，书中文责自负。

本书由上海同济城市规划设计研究院科研基金资助。

田莉 李晴
2015.10

第一部分
课程简介
Introduction to Course

A Brief Introduction of the Urban Planning and Design Course for International Students
面向国际留学生的城市规划与设计全英语课程建设简介

田莉　Li Tian
李晴　Qing Li

The English Urban Planning and Design Course is set up to meet the needs of urban planning education and globalization. In response to the further development of economic globalization and the Ministry of Education's calling of "Education should be oriented to modernization, the world and the future", Chinese higher education institutions began developing general and specialized courses taught in English. As the first urban planning department in China, the CAUP, Tongji University has already initiated substantial global exchange activities and our academic degrees are magnetic to more and more international students. Since 2005, we've awarded our master degree in urban planning and design to over 200 international students. Driven by the huge demand of English design course, we began to teach English urban planning and design course, and we believe it can help us better meet the need of cultivating high quality talents for the society. Since entering WTO, China has been in urgent need of talents specializing in different fields and, particularly, talents with both expertise and good foreign language skills. This posed a new challenge to Chinese higher education sector: to cultivate talents that meets global needs. It has become a common understanding in domestic higher education sector that opening courses taught in English will provide both international and Chinese students with a great platform of communication. It is extremely meaningful in helping them to develop imagination, creativity and global view, broadening knowledge scope, and improving their hand-on planning capabilities. This is the background in which the CAUP of Tongji University started building its English course platform.

1.1 A Brief Introduction of the English Urban Planning and Design Course

In the autumn of 2010, based on the experiences of joint Urban Planning and Design Studio, the CAUP of Tongji University officially opened its English Urban Planning and Design Course, providing global students with double master degrees programs. The students are mainly the graduate students from Bauhaus University, TU Berlin, Brandenburg Technical University Cottbus in Germany, Georgia Tech, Virginia Tech, University of Hawaii and University of Colorado in the USA, National University of Singapore, Milan Polytechnic and University of Pavia in Italy, University Lyon and University of Strasbourg in France, Chalmers University of Technology in Sweden, European University of Madrid in Spain, Aalto University in Finland and those from other universities in the four continents (Figure 1.1). For Chinese student, this course is optional, and limited space is reserved for Chinese graduate students.

All the lectures are delivered by teachers who have both overseas study experience and local urban planning and design education experience. The framework of urban planning and design courses varies with different universities in different countries. Many European and American urban planning and design colleges do not require their graduate students to provide a bachelor degree of urban planning and design. Therefore, international students in CAUP are from diversified education backgrounds, e.g. architecture, landscape, geography, sociology, municipal engineering, and even law. On the one hand, such diversification brought new perspectives into the process of the site study and planning and design process. On the other hand, students without drawing training will have to team up with those who have drawing skills to complete their plans and the final expressions of their ideas.

1.2 The Curriculum Framework of the English Urban Planning and Design Course

1.2.1 Objectives

The main objectives of this course are as follows: to develop case studies of selected global cities for benchmarking Shanghai's urban design and development; to explore cutting-edge concept and technology of urban design, to evaluate existing conditions and to initiate future scenarios and design concepts of the site based on the following analyses: urban form and typology, density mapping, economic development indicators and environmental effects. It will help students to explore emerging approaches and tools for design and analysis of the project, and enhance their abilities to implement the developed methods of critical thinking for urban planning and design in the fast-growing Chinese cities.

1.2.2 Schedule of Progress and Contents

Usually it takes student 16-17 weeks to finish this Course. The schedule of progress consists of six key links in three major stages, i.e. Site Study, Concepts Development, Final Expression and Review. The six key links are: Site Survey, Group Site Analysis Report, Individual Proposal, Regroup and Group Planning, Interim Reply, Planning Improvement, Final Reply and Review(Figure 1.2).

图 1.1　国际学生的来源地分布
Figure1.1　Sources of international students

开设全英文城市规划与设计课程是规划教育和国际接轨的需要。随着经济全球化的不断深入，按照教育部"教育面向现代化、面向世界、面向未来"的要求，高等教育要创造条件使用英语等外语进行公共课和专业课教学。作为国内最早成立的同济大学城市规划专业，国际交流越来越频繁，前来攻读学位的国际学生数量越来越多。2005 年以来，已先后有超过 200 名国际留学生获得同济大学建筑城规学院硕士学位。在这种情况下，开设英文城市规划与设计课程成为必然的选择。其次，开设全英文城市规划与设计是培养高素质人才的需要。我国加入 WTO 以后，不仅需要大量的专业人才，而且需要既懂外语又懂专业的复合型人才，这就对我国的高等教育提出了一个新的任务：培养能适应国际需求的复合型人才。开设全英文课程，为中国学生提供了与国际学生交流的平台，同时对他们拓展想象力和国际视野、提升知识面和加强规划设计能力有积极的意义。同济大学的全英语课程平台建设正是基于此背景开展的。

1.1　全英文城市规划与设计课程建设概况

2010 年秋季，基于多年来同济大学国际城市设计工作坊的经验，同济大学城市规划系正式开设了面向国际双学位联培硕士生的全英语城市规划与设计课程。生源主要来自于德国包豪斯大学、柏林理工大学、科特布斯勃兰登堡工业大学、美国乔治亚理工大学、弗吉尼亚理工学院、夏威夷大学、科罗拉多大学、新加坡国立大学、意大利米兰理工大学、帕维亚理工大学、法国里昂大学、斯特拉斯堡大学、瑞典查尔姆斯理工大学、西班牙马德里欧洲大学、芬兰阿尔托大学等大学的双学位联培研究生，研究生的国籍涵盖四大洲共 15 个国家（图 1.1）。同济大学城市规划系的研究生则采用自愿选修的原则。负责教学的老师均为有海外大学教育和本土规划设计教育双重背景的任课教师。

由于国内外城市规划教育课程设置的差别，欧美的城市规划专业硕士研究生并不要求学生具有设计的本科训练，因此国际留学生的本科教育背景相对多元化，如建筑学、景观学、地理学、社会学、市政工程乃至法律等，这会为基地的解读和方案构思注入新的视角，但部分同学由于缺乏规划设计的绘图技能训练，必须和其他有设计背景的学生共同组成设计小组合作进行方案构思与成果表达。

1.2　全英文城市规划与设计课程建设框架

1.2.1　教学目标

本课程的主要目标是以世界主要城市的经典城市设计为参考案例，与中国上海（或其他城市）进行比较，探讨基地的发展定位、机遇与挑战；探讨前沿的城市设计观念与技术在规划设计学科的应用。从空间形态、人流活动及功能定位等几方面来研究地区的现况，并形成城市设计的思路与城市设计概念。本课程促使学生探讨前沿的规划设计方法和工具，思考面向快速发展的中国城市设计的方法及其应用。

1.2.2　教学进度及内容

本课程时间共计 16~17 周，课程设置主要包括三个阶段共六个环节。三个阶段为：基地观察及分析、设计概念生成和成果制作及最后评图，六个环节依次是基地踏勘调查→小组基地分析报告→个人方案构思→根据方案重新分组，集思广益完成小组方案，进行中期答辩→修改、完善并深化规划设计→期终答辩及评图（图 1.2）。

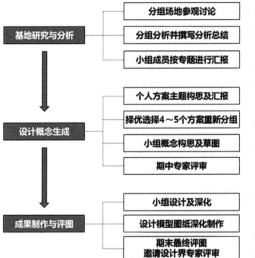

图 1.2　全英文城市规划与设计课程进度安排

1.3 Teaching Approach of the English Urban Planning and Design Course

1.3.1 Participation into real projects

During the Course, students shall participate in real planning projects. On the CAUP platform, the teachers may select for students some real urban planning and design projects from clients, such as Shanghai Municipal Planning and Land Resources Administration (SMPLRA)and its local branches, e.g. Jinshan District PLRA, Baoshan District PLRA, etc. after consulting with clients, we can select a project which the client is interested in, but does not have specific requirements. By partially involving students in real projects, we are able to avoid the inconsistence between the project progress plan and our course schedule. Up to now, our students have already participated into many real projects. For example, in 2010, they took part in macroscopic planning projects such as Thematic Plan of 2040 Shanghai Strategic Planning', mesoscopic planning projects (50-80 hectares) such as Urban Design of Coastal Area in Shanghai Jinshan District, 'Urban Renewal of the Old Town of Tinglin 'Urban Design of Baoshan International Cruise Terminal Area', as well as microscopic planning projects (10 hectares) like 'Civic Park Design of Jinshan new city'. Motivated by the kindly supports from the client, particularly, the actions it took to help students understand the actual operation modes of local urban planning authorities, our students showed great sense of responsibility toward the projects and great enthusiasm for study. The real-project-based teaching received a warm welcome from both the client and students. On the one hand, the students felt no restrictions on their creativity. On the other hand, the client is looking forward to ideas that could 'light up their eyes'. As a matter of fact, thanks to the inputs from the clients and the restriction of actual project circumstances, the students received a good training in rational analysis and the ignorance of actual local conditions was avoided.

1.3.2 Individual work and group work

The study contains two integral parts: individual work, which is also the essential part of student performance review, and group work, by way of which the final results are presented. In Stage I, i.e. Site Study and Analysis, considering the area of the projects they participate in is large, we'd send groups of 4-5 students coming from different backgrounds to study the site and explore its essence from social, economic, special and cultural perspectives under the guidance of instructors, and then, let every student present in group what they learnt from the perspective of his/her expertise. In Stage II, Concepts Development, the students shall work individually to fully develop and present their design concepts in sketches or in other ways such as brainstorming. Such freedom is provided to encourage students without urban planning training to actively take part in the planning and design process. No idea shall be judged. All ideas are recorded as a reference to 'spark off' the thinking and creativity in others' mind.

Based on the results of individual sketch concept presentation and discussion, instructors shall select 4 or 5 promising concepts and encourage students to regroup under the concept that interests them most. Then, the instructors shall fine-tune the structure of every group in accordance with the expertise and home countries for students to make the best of the diversified cultural background and learn from each other. The many aspects involved in urban planning dictated that it is a process of cooperating with people with different expertise, from the complicated site survey at the very beginning, through questionnaire investigation, subject-oriented study, to space design, and until overall planning and detailed planningogress and assign design tasks to members according to their expertise and interests, the instructors shall mainly play the role of inter-group progress coordination, design review and evaluation, and provide them with timely instructions and supports (Fang et al, 2012).

1.3 全英文城市规划与设计课程教学方法

1.3.1 选题与项目实际需求相结合

本课程采取"真题假作"的方式。借助同济大学建筑城规学院的平台，任课教师通过和上海市规划和国土资源管理局及各区县的规划和国土资源管理局（如金山区、宝山区等相关机构）的沟通协助，选择他们有实际研究和设计需求、但对设计成果无限定要求的项目。这样一方面得到来自甲方的积极支持，另一方面，又避免了如果开展设计的是"真题"，导致项目进度与教学进度不一致的情况。2010年以来开展的题目既有如"上海2040"这样的畅想型宏观层次的题目，也有上海金山区滨海城市生活岸线地区城市设计、亭林镇老镇区改造更新规划、宝山国际邮轮港滨江地区城市设计这样空间尺度上介于50~80公顷的中观层次的题目，还有上海金山区中心区核心开放空间设计（约10公顷）这样微观层次的题目。由于得到甲方的支持，并且帮助留学生了解地方规划管理机构的运作情况，可以激发学生的责任感和学习兴趣。同时，因为采取了"真题假作"的形式，一方面学生的创新能力不会受到限制，甲方也期待看到令人"眼前一亮"的想法；另一方面，学生的理性分析能力得以锻炼，因为有甲方的参与和实际情况的制约，可避免出现方案过于天马行空，不考虑地方实际的情况。

1.3.2 教学形式采用个人独立工作与小组合作的方式

由于设计课程的最终成果以小组形式呈现，但最终考核以每个同学的表现为基础，所以在教学环节采取了独立工作和小组工作相结合的形式进行。考虑到规划设计涉及的基地范围面积较大，在第一阶段的基地调研与分析阶段，往往由教师引导，不同专业背景的同学（4~5名）组成小组，对基地从社会、经济、空间、人文等不同的视角进行调研分析，形成对基地的基本了解，在做小组汇报时就自己所做部分进行汇报。在设计概念生成阶段，则采用独立工作的形式，可以各种方式（如头脑风暴式的畅想）展现个人的设计理念，不局限于图纸，这一方面兼顾了非设计背景同学的需要，另一方面也鼓励他们积极参与规划设计过程。任何创意不会受人批评，所有灵感均记录以备参考，使每个人的思考与创意成为启发别人的"火花"（范霄鹏等，2011）。

基于每位同学提供的创意，教师再从中选择4~5个具有潜力和个性的创意，鼓励学生选择自己感兴趣的概念重新分组。在重新分组的过程中，任课教师也会有意识地引导学生根据不同专业背景、不同国别进行分组，以充分利用跨文化背景下互相学习的优势，取人所长，补己之短。城市规划设计内容决定了设计工作是需要多工种合作完成的工作。从复杂的现场调研、问卷调查到专题研究、空间设计，从总体规划到细部设计，都需要多人共同配合完成。这就要求学生必须具备团队精神和合作意识。在工作室中，学生可以设计小组为单位，针对设计工作中的不同环节，进行分工完成。在学习中形成互相交流、互相协调、互相融合的意识。由学生自由选出设计组长；组长制定设计任务进度表，依据组员的特长和意愿分配具体设计任务；教师总体把关，协调并控制每个设计小组的设计进度并及时予以指导和帮助（方茂青等，2012）。

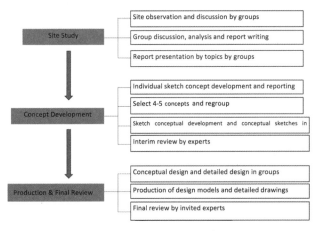

Figure1.2　Schedule of Progress

1.3.3 Combination of heuristic and situational teaching

The teaching also includes two integral parts: heuristic teaching and situational teaching. The Heuristic teaching is a pedagogy proven to be effective in developing the urge in students to take the initiative in making innovations and exploring the unknown by focusing the teaching activities on improving such capabilities as independent thinking, innovation and hand-on skills. On the other hand, the situational teaching tries to intrigue students by creating intuitive situations with possible resources and lead their curiosity to key learning points, learning methods and important content as a whole. This course gave up the traditional teaching mode in which instructors show students how to improve his/her design and drawing skills one by one and turned to a teaching approach combining heuristic and situational teaching. Specifically speaking, in this course, real urban planning projects and classic sites and cases from Shanghai and surrounding areas are used for students from different cultural backgrounds to explore how to apply different urban planning concepts and technologies for projects in the fast-developing Chinese cities, so that they could have a better understanding over the practices, characteristics and development of urban planning and design in China. In terms of teaching contents, this course pays emphasis on both the planning and design of physical forms and the social, political, economic and cultural circumstances behind them, to help international students understand how and why such physical forms came into being and put forward space plans that fit local development needs. In addition, the course tries to introduce the latest urban planning and design concepts in the world and help the students better use various skills to analyze specific planning problems.

1.3.4 Whole process participation of the Client and invited experts

The students have a number of opportunities to present concepts and plans individually or in group during the planning and design process of a project. Since each project we selected represents a real problem to be resolved by the client or a community, the client shall take part in the planning and planning evaluation at every key project stage. But, at the time of middle-term and final assessment of student academic performance, the CAUP shall arrange other facilities and invite some urban designers from planning and design firms with rich experiences in this field to review students' works. These comments are valuable to students and are expected to enhance the interest and confidence in them in urban planning and design.

1.4 Conclusion

For more than four years, we've been improving the contents and teaching approaches of this English Urban Planning and Design Course under the double master degree course and our practices and exploration were well-received by both international and Chinese students. Such positive response owes not only to the carefully designed course framework and teaching concepts, but also to the delightful surprise originating from the combination of diversified cultures and local characteristics. Specifically speaking, the students benefited from the course in three ways: the development of global view, the improvement of professional skills, and the communication and exchanges between international and Chinese students. In terms of the global view, in a time when China is accelerating its urban development, some very interesting urban planning and design issues emerged in front of urban planners. Some of them are problems students from well-developed countries have never met before. The freshness and challenges the international students felt on the land of China broadened their vision in a great way, let alone the improvement of the capability of observing planning issues in different ways. In terms of the improvement of professional skills, although the scope and objectives of study and the specific problems are entirely different from what the international students previously experienced, the course devoted a great deal to the essential knowledge about the urban planning and design and put great emphasis on guiding students to methodologies to help the students to find out the underlying factors and driving forces behind the complicated circumstances and explore frontier approaches in the urban planning world. The communication and exchanges between international and Chinese students is beneficial to both sides. The debate and discussion between international students and Chinese students, who are a smaller part in the program but contributed a great deal to the whole group, e.g. bringing in group discussion their understanding of the current situation in China and Chinese culture, a fresh perspective over specific problems, as well as their efforts in collecting Chinese materials, effectively helped both sides to see through superficial phenomenon and set their eyes on underlying factors. In short, through such debates and communications, overseas and Chinese students not only improved mutual understanding and built good relationships with each other. Most importantly, they are able to develop better design concepts and promote the integration of diversified cultures and local characteristics.

Reference
Fan, Xiaopeng and Yan, Huimin Research on the Contents of Staged Urban Planning Courses, Planners[J]. 2011(27), 263-266.
Fang, Maoqing and Tian, Mimi, Thoughts on the Teaching of Urban Planning Course in the Context of Studio-based Planning Service Mode, Central China Construction[J]. 2012 (5), 167-170.

1.3.3 教学手段采取启发式和情景式教学相结合的方式

教学方法上，主要采用启发式教育方法和情景式相结合的方法。在培养学生创新意识、探索精神、创新能力方面，启发式教学法是一种有效的方法。它把发展学生独立思维能力、培养学生创新能力和实践能力作为教学的核心内容。情景式教学是在教学中充分利用条件创设具体生动的场景，激起学生的学习兴趣，从而引导他们从整体上理解和把握教学中的重难点、学习方法和关键内容。本课程在教学方法上，突破以往规划设计课程老师手把手改图的传统教学方法，应用启发式和情景式教学，选择上海及其周边城市真实的设计题目和具有代表性的场地与实例，探讨在快速发展的中国城市中，城市规划设计观念与技术如何应对城市发展，使不同文化背景下的学生全面深入了解中国城市规划设计发展的实践和特点。教学内容上，不仅关注物质性的形态规划，而且关注形态背后的社会、政治、经济和文化机制，以帮助国际学生理解空间形态的生成机制并提出适应地方发展的空间方案。此外，通过对国际上前沿规划设计理念的介绍，帮助学生运用相关技能分析规划的相关问题。

1.3.4 评价上采取甲方全过程参与和专家参与相结合的方式

在城市规划设计过程中，学生有多次汇报独立概念和小组方案的机会，由于选择的是带有甲方需求的小区的设计题目，所以在每个关键的环节都有甲方的参与和评审。而在中期考核和最终考核阶段，则往往邀请校内其他规划设计教师和设计公司富有实践经验的著名设计师进行点评，让学生可以从评价的各个环节都能有所收获，并提升对设计的兴趣与信心。

1.4 总结

经过四年多的探索与实践，国际双学位联培硕士生的全英语城市规划与设计课程的教学内容和教学方式已趋完善，受到留学生和本国学生的欢迎。这些积极的反馈既是基于本课程较为严谨的教学框架和教学理念，也源自于跨文化与本土性结合带来的惊喜，具体表现在以下三方面：第一，培养学生具备全球视野，在我国快速城市化的背景下，出现了许多有趣的规划设计和城市研究的课题，一些课题是西方发达国家的学生所未曾触及的，进入中国现场，这使得留学生获得一种从未有过的新鲜感和挑战性，开阔了留学生的视野，通过比较分析，增强综合判断和分析的能力；第二，提升学生的专业性知识，对于留学生而言，尽管研究对象的尺度和具体问题与之前大为不同，但本课程关注于城市规划和设计的知识本体，注重方法论的引导，让学生从复杂的表象背后挖掘潜在的影响因素和动力机制，探索前沿性知识和规划设计方法；第三，中外学生之间的交流，让彼此获益，尽管中国学生在人数上较少，但是在小组讨论中，中国学生对中国的现实和文化理解，对中文资料的收集，对不同问题的看法，以及中外学生之间反复的辩论研讨，使得许多问题能够透过表面现象，深入理解，这种辩论让中外学生双方相互理解，加深了感情，也深化了方案的概念性，有助于跨文化与本土性相结合。

范霄鹏，严佳敏，城市规划专业设计课程阶段性教学内容研究，规划师，2011（27），263-266
方茂青 田密蜜，以工作室制模式为背景的城市规划设计课程的教学思考，华中建筑，2012（5），167-170

第二部分
教学方法、过程和成果要求
Teaching Methods、Process and Final Production

2.1 Urban Planning & Design for International Students: Site Study and Approaches
2.1 面向国际留学生的城市设计：基地分析与研究方法

田莉　Li Tian

Urban planning is a process revealing the urban functions of a site and the activities of people living there. It aims at consciously changing and improving the site to enable it provide people with suitable urban functions and meet their needs of carrying out various activities in accordance with proper principles. Also, the urban planning is a process of constant communication between the planner and the site. In the conversation between the planner and the site, the planner is first of all a listener (Zhang, 2012) who wants to thoroughly understand the characteristics, limitations and potentials of the site from different perspectives: the time, i.e. history, present and future, and the space, i.e. the physical space of the site and surrounding areas, and even broader perspectives like culture, society and economy. The ultimate goal of urban planning is to achieve the Harmony between Man and Nature. The Nature refers to the natural environment, i.e. terrain, landforms, climate, vegetation system, current environment, surrounding environment, history, culture, etc. The Man refers to the people living on the site, their need of conducting different activities, and the reasons behind them, i.e. social, economic, cultural and living and so on. At the site study stage, the planner has to thoroughly study all factors with respect to the Nature and Man to be able to deliver the Harmony on the next stage, concept development. Only in this way can a planner develop a relevant plan reflecting the characteristics of the site, instead of playing a stereotyped 'game of mapping' (Zheng, 2007).

2.1.1 Levels and Contents of a Site Study
2.1.1.1 Site study: Space analysis

The space analysis includes three aspects: Macroscopic analysis, Mesoscopic analysis and Microscopic analysis. First, the macroscopic analysis usually studies the space factors at the level of a city or districts in a big city. For example, the macroscopic space analysis over the 'Urban Design of Coastal Area' in Jinshan District should focus on understanding the importance of the site to the origin, current status and future urban development in Shanghai, based on the unique resource the site could provide, i.e. Jinshanzui Village, the earliest and only existing fishing village in Shanghai, and what its seashore resources mean to the metropolis. Second, the mesoscopic analysis usually studies the space factors at regional level. Take the Jinshan Project as an example again. The focus should be placed on understanding the location of the site from the perspectives of regional traffic and transportation, ecological system and space structure and the essentials of master planning at the level of Jinshan District. Third, the microscopic space analysis usually studies the space factors of the site itself and surrounding areas, including natural conditions, current construction, and historical and cultural factors. The natural conditions analysis includes the regional environment, daylight, winds and other climate factors. The terrain analysis covers the elevation, slope, slope aspect, bearing capacity, etc. The hydrological analysis involves water body (water system), water quality, water quantity, water catchment lines, wetland, and underground water and so on. The ecological sensitivity analyses focus on determining whether an area is suitable for construction. Current construction status analysis aims at understanding the distribution of road and traffic facilities, public facilities, business and retail facilities, office buildings, tourism facilities on the site and in the surrounding areas, as well as the current status of constructions on the site, assessment of the style and appearance of buildings, historic communities and spaces with special features, and everything that may affect the future layout of the project. The historical and cultural analysis shall be centered on understanding the historical, economic and social characteristics of the site.

2.1.1.2 Site study: Time-dimensional analysis

From the perspective of time, the site study can be carried out at three levels, i.e. historical, present and future levels. It's impossible to do the right things at present and foresee the future without a proper understanding of the history. From the point of the site study, historic events at different eras shall always leave a mark on the physical space. A good urban planning shall never ignore the cultural history of a site. For example, the site of Jinshan coastal area urban design used to be lying under sea water and gradually developed into land as the earth brought by the Yangtze River accumulates (Figure 2.1). How to integrate historical elements into current urban plan is a question that should be considered in all urban planning projects. As time goes by, along with the emerging of new human activities, industries and functions, e.g. creative industry and network communities, people shall have different use of the same space. The core of urban planning and design, therefore, should cover both the present and the future, i.e. paying equal attention to current and future functional needs and developing forward-looking space layout.

2.1.1.3 Site study based on the analysis of human activities and urban functions

In the site study, from the perspective of populations who use the site, the population can be classified into two groups: local population (people with local household registration and permanent residents) and migrant population (commuters who work here but do not live here), as well as local tourists and other tourists, if the site has tourism resources. Different populations apparently have different feelings about the use of space and different needs of space. The importance of understanding these needs by means of investigations and interviews to the development of urban planning ideas should

城市设计是一个不断了解熟悉基地的城市功能与人们生活活动规律的过程，对基地进行一种有目的的改变和提升，使之能适应相应的城市功能与人们活动的需求与规律。设计过程是一个设计者与基地不断交流的过程。在人与基地间的对话中，人首先是一个倾听者（张东红，2012），需要从时间维度——历史、现在、未来，空间维度——基地自身、基地周边乃至更广大的区域范围及超越空间的人文、社会和经济视角，全面分析和辨别基地的特征、限制条件和发展潜力。城市设计的最高目标是"天人合一"。"天"是指自然环境（地形地貌、气候、植被、现状与发周边环境……）。"人"是指人们在基地上的活动需求与规律（社会、经济、文化、生活……）。在基地分析阶段，应将这两方面的要素研究透彻，才能在下一阶段设计概念的生成中使两者"合一"。唯有如此，才能做出切合基地特点的设计方案，而非千篇一律的"图形游戏"（郑正，2007）。

2.1.1 基地分析的层次和内容
2.1.1.1 基于空间维度的基地分析

从空间尺度上来分析，基地分析应该包括如下三个层面的内容：①宏观层次：一般来讲是市域或都市区层面。如金山城市生活岸线区域的区位分析，应从自身的资源入手：上海最早而且仅存的渔村——金山嘴渔村和海岸线资源对上海大都市区的意义入手，分析该片区对于上海城市发展的起源、现实和未来的重要意义；②中观层次：一般指区域层面。同样以金山区城市生活岸线为例，需要从金山区层面，就区域交通、生态系统、空间结构等方面来分析该片区所处的区位，明确上位规划的要求；③微观层次：指基地本身及周边地区，分析内容包括自然条件、现状建设状况和人文状况等。自然条件包括分析基地所在的区域环境、日照、风向等气候特征；地形分析包括高程、坡度和坡向，地质承载力等；水文分析包括水体（水系）、水质、水量、汇水线、湿地、地下水等多种要素。基于生态敏感性等分析，确定适宜建设和不适宜建设的区域。现状建设状况指基地及周边道路交通设施、公共服务设施、商业零售设施、商务办公设施、旅游设施等的分布状况及基地内现状建筑建设状况、建筑风貌的评价，历史街区和特色空间等，这些都会影响基地未来的布局。人文状况分析多包括区域的历史、经济、社会等特征。

2.1.1.2 基于时间维度的基地分析

从时间维度上来划分，基地分析的层次可以分为历史—现在—未来。不了解历史就无法把握现在和预知未来，在基地分析中，不同时代所历经的事件往往会在物质空间上留下痕迹，一个好的城市设计不能割断历史的文脉。如金山城市生活岸线地区历史上曾属于海域，随着海岸线的不断延伸而逐渐成为陆地（图2.1），如何从现有设计中寻找历史的遗迹就成为设计必须考虑的内容。随着时间的变迁，新的活动、产业和功能产生，如创意产业、网络社会等的出现，对空间使用产生了新的需求。在现在和未来两个维度上，着眼于现在和未来的功能需求，有前瞻性地进行空间布局，是城市规划与设计的核心内容。

2.1.1.3 基于人群活动需求和城市功能营造的基地分析

从基地的使用人群划分，基地分析通常需要着眼于本地人群（包括户籍人口和外来常住人口）和外来人口（通勤人口，在此工作但不居住于此地的人群）。对于有旅游资源的基地而言，还包括本市观光者和其他省市外来观光者。不同群体对空间使用的感受不同，对空间的需求各异，通过调查访谈了解他们的需要，对城市设计的概念生成具有重要意义。如金山城市生活岸线中，一个设计小组就针对不同的使用对象的特征和空间需求进行了深入分析，形成了生动有趣的"人物图谱"（图2.2）。

图 2.1 金山海岸线变化
Figure 2.1 Evolution of coastlines of Jinshan

never be underestimated. For example, in the urban deisgn of Jinshan coastal area, one design group developed a very interesting and lively 'Face Map' (Figure 2.2), based on a careful analysis over the characteristics and needs of different peoples who use the site space.

Every city/site has its own unique characteristics, functions and development position. It is necessary to analyze the site from the perspective of functional development. For example, in the urban design Baoshan International Cruise Terminal Area, the site study should have focused on its role in the self-positioning of Shanghai as the "Economic, financial, trade and shipping center". The difference between the current conditions of the site and its future status could be gradually covered and completed during specific urban planning and design.

2.1.2 Common Approaches of a Site Study
2.1.2.1 Information sources

The information materials necessary for a site study can be classified into two categories by sources: first-hand materials (or original materials) that obtained by hand-on collection and direct experience, and, second-hand materials (or secondary materials) that quoted from the existing materials or the re-interpretation of existing materials. The first-hand materials are valuable for their verifiability, vitality and readability. If planners can complete space dimension analysis simply by applying professional knowledge and skills, when it comes to the first-hand materials, they have to conduct a lot of site surveys and interviews to fully understand how the people living on the site feel about the space. In urban planning, the first-hand materials can be obtained by way of questionnaire and in-depth interview and presented in charts, video and audio records based on the results of the questionnaire. The secondary materials include master planning report, files, collective drawings, etc. The secondary materials are complementary to the original materials collected via on-site investigation and observation. A proper understanding over the functional spaces and the trend of future development of the site can only be achieved by combining both of them.

2.1.2.2 Questionnaires and in-depth interviews

The questionnaire survey is a research instrument used for gathering information from a particular group of respondents. It uses a series of questions and other prompts to gather the respondents' opinions, feelings and responses to reveal their perception of a space. Data can be collected by questionnaires or interviews or any other means. The questionnaire is adopted under the condition that the planner has already defined the questions he/she wants to ask. Then, the questions shall be presented in a table, printed, issued to and filled by a target respondent group, collected, and finally analyzed for a conclusion. On the other hand, the interview is an unstructured, immediate and face-to-face research instrument. During an interview, student participants shall have an in-depth discussion with the site users to understand their feelings about the site, the problems of the site in their eyes, etc. the questionnaire and in-depth interview are the most important instruments to directly sourcing first-hand materials.

2.1.2.3 Observation

Observation is one of the important activities in mankind scientific practices. The observation survey is one of the fundamental methods used in today's urban planning teaching. It is also an essential instrument for urban planners to have a personal experience of the site and understand the characteristics of human activities on the site. The observation in urban planning is a kind of nonparticipant observation during which a planner is irrelevant with the observation targets and should restrict himself/herself from affecting the behavior of the observation targets or any on-going activity, or participating into any activity. The role he/she plays is nothing but a recorder. The information collected through observation is truthful and reliable first-hand information.

The observation is usually carried out at different points on the site at different periods of time. For example, in case of an public space, we need to observe the human traffic, transportation, leisure activities, art performance, social activities and so on in the daytime and evening on a week day or holiday to see the activity patterns of the elderly, the young, the teenage, commuters, tourists, etc. What is the difference between week days and weekends, day and evening? What are the characteristics of activities on the site in different seasons, the characteristics of populations using the site at different time? The conclusion originated from the summary of the characteristics of activities happening at different periods of time may be presented in various ways. For example, in the Lujiazui Urban Regeneration Project, our students used an 'activity tree' to present the characteristics of human activities at different time in a day. (Figure 2.3)

图 2.2 人物图谱　图片来源：课程教学学生绘制
Figure2.2 "Face Maps" of Different people　Source:Students' drawing

每个城市/基地都有其自身的特色，亦有其自身的功能与发展定位。从城市功能营造的角度的出发，对基地进行分析十分必要。如进行宝山国际邮轮港滨水地区的基地分析时，就应着眼于该区在上海打造"四个中心"（经济、金融、贸易、航运中心）的城市功能定位中的角色，分析基地现状和预期目标之间的差异，在基地规划设计中予以补充和完善。

2.1.2　基地分析与研究的常见方法

2.1.2.1　基地分析的资料来源

从资料来源来划分，基地分析所获取的资料主要包括两部分。一为第一手资料（又称原始资料），指自己直接经过搜集整理和直接体验分析所得。第一手资料具有实证性、生动性和可读性的优点。如果说基于空间尺度的分析主要基于设计师所受到的专业知识技能经过分析而获得，第一手资料则主要指设计师对现场的调研和访谈获得的资料，借此深入了解生活在基地上的人对空间的切身感受。在城市设计中，可以通过问卷调查，深度访谈等形式获取第一手资料，资料的表达方式包括问卷调查后的图表分析、视频录像、声音记录等。二为第二手资料（又称次级资料），是指对于已有资料的引用与再诊释，如上位规划报告、文件、图集等。它与实地调查法、观察法等收集原始资料的方法是相互依存、相互补充的。只有将第一手资料和第二手资料结合起来，才能较为全面地了解城市设计中需要完善的功能空间及基地未来发展的趋势。

2.1.2.2　问卷调查与深度访谈

问卷调查是有目标对象群的意见调查方法。形式是由一连串写好的小问题组成，收集被访问者的意见、感受、反应及对空间的认识等。设计师可以用问卷调查收集数据，也可以用访谈或其他方式收集数据。问卷调查假定设计师已经确定所要问的问题。这些问题被打印在问卷上，编制成书面的问题表格，交由调查对象填写，然后收回整理分析，从而得出结论。访谈法是一种无结构的、直接的、个人的访问，在访问过程中，参与设计的学生对场地的使用者进行深入访谈，以了解其对场地的感受、场地存在的问题等。问卷调查与深度访谈是获取第一手资料的重要和直接的来源。

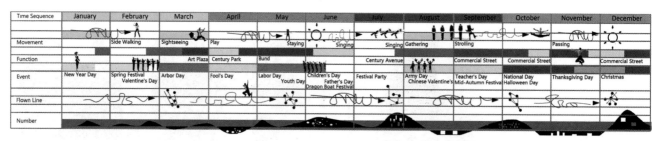

图 2.3 陆家嘴金融中心区在一天不同时段的人群活动特征
图片来源：课程教学学生绘制
Figure 2.3 "Activity Trees" of Lu Jiazui Residents
Source: Students' drawing

2.1.2.4 SWOT Analysis and other instruments

SWOT (Strengths/Weaknesses/Opportunities/Threats) analysis was first developed to help businesses identify their strengths, weaknesses, opportunities and threats and, thereby, effectively align company strategies with internal resources and external environment. By analyzing specific situations, the SWOT can help a planner to identify and evaluate various internal strengths and weaknesses and external opportunities and threats relevant to a project, and then clearly present them in a matrix. After that, it can help the planner to conduct systematic analyses over all factors presented in different combinations and, eventually, build effective strategies (Xiao, 2009). As the SWOT enables its user to effectively combine the internal and external factors, it is recognized as a comprehensive systematic analysis instrument featuring good application adaptability and intuitive presentation of results. Considering the fact that urban planning is a comprehensive process involving so many factors, e.g. economic, geographic, social, cultural, technological factors, to name a few, applying SWOT analysis in the site study can help planners to have a clearer view of the site's strengths, weaknesses, as well as the challenges before it (Figure 2.4).

Of course, there are other analysis instruments available to site study, including those are more effective in analyzing cases involving too many factors, e.g. principal component analysis, quantitative analysis type I, II and III, etc (Zhang, 2000), to name a few. Planners may choose from them according to the specific circumstances of the site.

2.1.3 Site Study Results

The results of a site study are usually presented in drawings with brief text description based on the objectives and specific client requirements of the site study, including:

2.1.3.1 Drawings

(1) Location analysis Chart

In general, the Location analysis Chart (LAC) presents location analysis at three levels. Macroscopic LAC shows the location of the site at city/city district level, identifying the exact location of the site in the city and its distance from the city center, key transportation facilities such as metro stations and harbors, and demonstrating the accessibility of the site. Mesoscopic LAC should clearly identify the locations of district center, key district transportation facilities, and district public facilities, and give a more detailed description of the accessibility of the site at the district level. Microscopic LAC should clearly demonstrate the distribution of roads, traffic system, public service facilities on the site and surrounding areas.

(2) Status quo Analysis Charts

• Current Land-use Chart: showing the boundaries and purposes of different lands on the site in accordance with the Standards for Classification of Urban Land and for Planning of Constructional Land, 2012;

• Building Quality Assessment Chart: showing the construction status, particularly the construction structure and quality, of existing buildings on the site; There are several types of building quality assessment chart available, providing a reliable reference for identifying buildings to be kept or demolished;

图 2.4 SWOT 分析法　图片来源：课程教学学生绘制
Figure2.4　SWOT Analysis　Source:Students' drawing

2.1.2.3　场地与活动观察法

观察是人类科学认识中的重要实践活动。观察法在城市设计教学中是一种最基本的方法，它是设计师获取感性经验和了解场地活动特性的根本途径。在城市设计的观察过程中，常采用非参与性观察的方式，即设计师与被观察对象无关，对于被观察对象的行为与事件的发展不施加任何影响，也不参与任何活动，而只是充当记录者的角色，所获得的信息资料，具有真实可靠性，是第一手资料。

在进行活动观察时，需要对场地的不同节点，在不同时段内进行观察。例如，如公共活动空间平日、夜间、节日的人流、交通、休憩、演出、活动等，老人、上班族、青少年、游客等的活动规律。平时和周末的场地活动有何不同？白天与傍晚的活动有何不同？不同季节的场地活动有什么样的特性？不同时段使用场地的人群有何特征？总结特定时段活动的特征后可以以各种形式来表述。例如，在陆家嘴城市更新设计中，同学以"矩阵"的形式来表现陆家嘴金融中心区在一天不同时段的人群活动特征（图 2.3）。

2.1.2.4　SWOT 分析法及其他方法

SWOT（Strengths Weaknesses Opportunities Threats）分析法，又称态势分析法或优劣势分析法，最初用于确定企业自身的竞争优势（Strength）、竞争劣势（Weaknesses）、机会（Opportunities）和威胁（Threat），从而将公司的战略与公司内部资源、外部环境有机地结合起来。SWOT 分析法通过具体的情景分析，将与项目密切相关的各种内部优势因素、劣势因素和外部机遇因素、威胁因素分别识别和评估出来，依据矩阵的形态进行科学的排列组合，然后运用系统分析的研究方法将各种要素相互匹配进行分析，最后提出相应对策的方法（肖鹏飞等，2009）。SWOT 分析将内外要素有效结合，是一种全面、系统的分析方法，其优势在于方法运用的灵活性和表达成果的直观性。由于城市规划设计涉及经济、地理、社会、文化、技术等诸多因素，具有综合性强的特点，在基地分析中采用 SWOT 分析，对明确基地的优劣势和面临的挑战等具有条理清晰、明确易懂等特点（图 2.4）。

当然，在基地的分析方法中，针对影响因素较多，还包括主成分分析法，数量化 I 类、II 类、III 类方法等（章俊华，2000）。根据基地的不同，可以分别进行选择，篇幅所限，不一一列举。

2.1.3　基地分析的成果构成

结合基地分析内容，根据具体项目及设计要求，基地分析研究的成果以图纸为主，结合文字说明。包括以下内容：

2.1.3.1　图纸构成

（1）区位分析图

一般情况下，区位图包括三个层次，宏观的都市区 / 市域层面的区位图，需要标明基地在都市区的位置，并标注距市中心、重点交通设施如机场、高铁站、港口等的距离，以说明基地的可达性。中观层次的区位分析图应包括该区域内的中心区、区级重要的交通设施、公共服务设施等的位置，在次区域层面上对基地的可达性做进一步诠释。在微观层次的区位分析图上，需要标注基地周边地区的道路交通状况、公共服务设施布局状况等。

（2）基地现状分析图

• 土地使用现状图，需要参考住房和城乡建设部的《城市用地分类与规划建设用地标准》GB 50137—2011，对基地内现有的各类用地的界限和使用性质进行说明。

• 建筑质量评价图：对基地内现存的各类建筑的建设状况，主要是建筑结构和质量进行评价，可以分为若干类，为确定哪些建筑需要保留、哪些建筑可以保留、哪些建筑可以拆除提供依据。

• Functional Analysis Chart: showing and analyzing the functions of lands and buildings on the site to identify key functions and secondary functions;

• Public Facility Distribution Chart: Identifying the public service facilities on the site, i.e. education, culture, public health and sports facilities;

• Current Roads and Transportation Chart: showing main roads, secondary roads, by-passes, parking lots, urban railway stations, bus lines, etc.

• Current Landscape Analysis Chart: showing and analyzing the landscapes on the site, e.g. iconic buildings, structures, parks, landscape nodes, lines of sight traverse, etc, providing an intuitive reference for planners to identify elements to be used or improved;

In addition to the drawings mentioned above, other analysis charts are also available for the specific characteristics of the site, e.g. internal and external water system, landscaping and vegetation systems, population activity patterns and demands, historical evolution of the site, the site's connection with the city in terms of culture, living and customs, etc. If necessary, planners can provide other materials like photos and video footages, as a complement.

Based on the contents of the drawings, they can be presented in two ways: concrete presentation and abstract presentation. For example, current land use, building quality assessment, public facilities are usually presented in a concrete way. On the other hand, analysis on the functional structure and landscape may be presented in abstract languages like iconic symbols, e.g. bubble chart, cycle diagram.

2.1.3.2 Text description

In the site study of an urban planning, the text description may be provided separately or together with the drawings to achieve better effects.

Reference:
Zhang Donghong, Site Analysis Approaches in Urban Planning and Design, Shanxi Architecture, 22-24
Zhang Junhua, Quantification Theory Type III: an investigation instrument in urban planning, Chinese Gardens, 16 (68) 2000 (2);
Zheng Zheng, Looking for Right Urban Planning for China, Zheng Zheng Urban Planning and City Planning Essays, Tongji University Publishing House, 2007;

- 功能分区分析图：对基地内的土地使用和建筑功能进行分析，确定哪些是主要功能，哪些是辅助功能。
- 公共服务设施分布现状图：对基地内的教育、文化、医疗卫生、体育等公共服务设施进行标注。
- 道路交通现状分析图：对基地内的主要道路、次要道路、支路、停车设施、轨道交通站点、公交线路等进行标注。
- 空间景观现状分析图：分析基地内的景观要素，如标志性的建筑物、构筑物、公园等，景观节点，视线通廊等，为城市设计中可以利用哪些要素，需要改善哪些要素提供依据。

除上述图纸外，可以根据基地的特点，提供其他分析图。如基地内外水系、绿化、植被系统等。此外，人群活动的规律、需求，基地历史演变的分析图、基地所在城市的相关文化、生活特征与习俗的文字与图示都不可或缺，并可提供照片、视频录像等作为补充说明材料。

在分析图的表达方式上，根据图纸内容的不同，分为具象和抽象两种表达方式：如对于土地使用现状图、建筑质量评价图、公共服务设施布局图等，一般需要采用具象的表达方式；而功能结构分析、景观分析等则可以采用象征和抽象的图解语言、形式来概括表达图纸内容，如泡泡图、循环图等。

2.1.3.2 文字说明

在城市设计的基地分析中，可以提供单独的说明书，也可以和图纸结合在一起，以达到图文并茂的效果。

参考文献：
[1] 张东红. 城市规划设计中的基地分析方法 [J]. 山西建筑，2012.07：22-24.
[2] 章俊华. 规划设计学中的调查分析法 10—数量化Ⅲ类 (Quantification Theory Type Ⅲ)[J]. 中国园林，2000，02：16-80.
[3] 郑正. 寻找适合中国的城市设计 [J]. 城市规划学刊，2007，02：95-99.

2.2 Generation Framework of Urban Planning and Design Concepts
2.2 城市规划设计概念生成框架

李晴　Qing Li

Urban planning and design concepts are innovative kernels of planning and design work. A good design concept is like a seed, which contains innovative DNA. By virtue of this gene, through logical deduction and creative expression, it is possible to achieve an innovative planning and design work. In order to obtain deliberate, unique and innovation-oriented design concepts, it is necessary to perceive and analyze the site. After accessing a certain design concept, it is also needed to further discriminate and interpret the effectiveness of such a design concept. As a result, the generation framework of urban planning and design concepts can be expressed as: Preliminary Study, Field Research, Eidetic Universal, Preliminary Concept, Conceptual Analysis, which are each respectively described below.

2.2.1 Preliminary Study

Preliminary Study refers to the collection, analysis and summary of information on planning and design objects before entering into the scene, which is the preparatory work of design concept thinking. In this stage, students are required to explore properties and traits of the site at macroscopic, mesoscopic and microscopic levels, analyze social, economic and cultural factors associated with the site, and probe into the potential of the site.

Take a project such as the one in Jinshan Coastal Area Urban Design as an example. This project is located in the southern part of Jinshan New Town and faces Hangzhou Bay. It covers rare coastal resources of a living shoreline and boasts rich historical, cultural and ecological resources. Therefore, it is necessary to analyze three dimensions of time, space and ecological environment, and properly handle the relations among industry, culture, landscape and ecology associated with the waterfront.

2.2.2 Field Research

Upon the completion of the preliminary study, on-site perception and intuition are a must. According to the famous U.S. architect Holl (1994), "Design ideas and concepts germinate when people perceive some places. In a work as a perfect combination of building and place, people can understand the meaning of place, implications of natural environment, real scenarios and feelings of human life, as well as harmony among artifacts, the nature and human life." The so-called "perception of the people present" carries and guides the thoughts and actions of the people present, which can immerse the people present into uncoveredness (Unverborgenheit) of perceived objects to make genuine and open-ended judgments (Tong Ming, 2009).

Maurice Merleau-Ponty (2001) considers it necessary to return to the "perceived life world," "Empiricism" and "Intellectualism" are two kinds of prejudice which hinder this return. In the course of on-site observation and description, just any scientific explanation should not be adopted. It is necessary to probe into the inherent diversity of a site. In doing so, designers might discover particularly interesting points from sterotypes.

In the Jinshan District (Shanghai) project, a student observed the repeated appearances of "water" in the site and expressed their own experiences and emotions in the form of color: "Organic form" of status-quo fishing village closely associated with "water", "cloud-enveloped" Jinshan Islands, etc (Figures 2.5 and 2.6). These appearances reflect some traits of the site and perception of these traits will lay an important foundation for subsequent design concepts.

During field research, it is necessary to communicate with the users as much as possible, as the users' perception and intention can also reflect some (potential) features of the places.

2.2.3 Eidetic Universal

Like the previous ones, this stage also requires putting aside any preconceived notions (e.g., any scientific theories judging on the objects, common sense and personal interests) so as to ensure that an open-minded attitude is taken towards the design concept. Without this process, we cannot make an accurate description of an object based on our immediate perception (Miao Pu, 2009). For example, there may be a "bias" towards the Chinese traditional folk architecture theory of "white walls and black tile (fen qiang dai wa)." Through on site observation, it is found that "fen qiang dai wa" is actually composed of a variety of different gray colors rather than white and black colors. In order to obtain eidetic intuition for essence of object, two key tasks must be fulfilled in this stage: intentional reduction and access to eidetic universal.

E. Edmund Husserl develops two reduction paths, i.e., the ontological reduction and the Cartesian reduction. An ontological reduction refers to shifting into a type of new reflective gesture from scientific exploration and thinking about a variety of intentionality. For example, for the Jinshan site, it is advised to constantly examine object-based ontology: What does a living shoreline actually mean? What does "urban" living shoreline mean? What does southern sea mean (for fishermen, tourists, Jinshan or Shanghai)? What does fishing village (with time-honored history) mean? To answer these questions, meditation, association and imagination are required so as to roll out a variety of new cognition and understanding and form the basis for the design concept.

A Cartesian reduction path is "trying to doubt" over various intentions rather than use Cartesian universal skepticism. For example, the common practice of reclamation zones can be questioned in many different ways:

城市规划设计概念是规划设计作品的创新内核。一个好的设计概念就像一颗种子，蕴含创新性的基因——DNA，凭借这个基因，通过逻辑演绎和形态表达，才可能实现一个富有创新性的规划设计作品。要获得一种缜密独特和创新导向的设计概念，需要对基地及文脉进行感知和分析；其次，在获得某个设计概念之后，还需要展开进一步的辨析和阐释，以确认设计概念有效性。由此，城市规划设计概念的生成框架可以表述为：初步研究→现场调研→本质共相→初步概念→概念解析，下面分别进行说明。

2.2.1 初步研究

初步研究是在进入现场之前，对规划设计对象的资料收集和分析归纳，是设计概念思考的前期准备，要求学生对基地的属性和特质预先进行认知，包括从宏观、中观和微观三个层面，分析与基地相关联的社会、经济和文化背景，分析基地的历史演变，考察全球典型案例，探讨基地的基质和潜力。

以上海金山区项目为例（见第76-105页），此项目位于金山新城的南部，面向杭州湾，具有稀缺的滨海生活岸线资源，基地内历史人文资源和自然生态资源丰富。在缺席意向层面，需要把控时间、空间和生态环境等三个维度，处理好基地与滨水相关的产业、文化、景观和生态之间的关系。

2.2.2 现场调研

在完成前期初步研究之后，进入现场进行感知和直观。现场感知非常重要，美国著名建筑师霍尔（1994）认为设计思想和概念从感受场所时开始孕育，在一个建筑与场所完美结合的作品中，人们可以体会场所的意义、自然环境的意味、人类生活的真实情境和感受，以及人造物、自然与人类生活的和谐。那种"在场者的知觉"，承担和引导着在场者的思考与行动，从而能够将在场者带入所知觉对象的无蔽状态，去从事那种真正的、开启式的判断（童明，2009）。

莫里斯·梅洛-庞蒂（Maurice Merleau-Ponty, 2001）认为需要回到"被感知的生活世界"，而"经验主义"（Empiricism）和"唯理智论"（Entellectualism）是阻碍这种回归的两种偏见。在现场观察和描述的过程中应该不带任何科学的解释和加减，不带任何偏见，把所有知识加入"括号"，切身感知场地，直观对象所固有的多样性，包括不同侧面、视角面、外形和属性，寻找基地特别有意思之处，从大家熟视无睹中发现某种闪光点。

图2.5 与水关联密切的基地现状渔村的"有机形态"
图片来源：课程教学学生绘制
Figure 2.5 "Organic Form" of status-quo fishing village closely associated with water
Source: students' drawing

在上海金山区项目中，某位同学观察到基地内反复出现的"水"的显像，并以色彩的方式将自己的体验和情绪进行表达：与水关联密切的"有机形态"的现状渔村和"烟云笼罩"的金山三岛等（图2.5和图2.6）。这些显像折射出基地的某种特质，对这些特质的直观将构成后来设计概念的重要基础。

作为一个规划设计项目，在现场调研阶段，还应该尽可能与使用者沟通，使用者对于基地的感知和意向也能反映出某种（潜在的）场所特征。

2.2.3 本质共相

与前一阶段一样，本阶段也需要搁置任何成见，包括任何对该对象做出判断的科学理论、常识和个人利害，以保证我们敞开胸怀地接受任何在意识中直觉到的显像本身……没有这个还原，我们就无法按第一时间所体验到的现象来描述一个对象（谬朴，2009）。例如，关于中国传统民间建筑"粉墙黛瓦"的理论可能是一种"偏见"，因为通过现场实地观察，可以发现"粉墙黛瓦"并非雪白和漆黑，而是由各种不同的灰色所组成。为了直观到对象的本质，本阶段需要从事两件重要的工作：意向还原和获取本质共相。

图2.6 天色将暗之时基地南面处于"烟云笼罩"之中的金山三岛
图片来源：课程教学学生绘制
Figure 2.6 "Cloud-enveloped" Jinshan Islands in the South of the Site at Twilight
Source: students' drawing

"Does forming a reclamation zone mean completely filling the sea with soil?" "Is the ancient fishing village certain to be replaced by commercial developments with a high floor area ratio?" "Will reclamation zones be wholly linked by roads (or waterways)?" "Will buildings in reclamation zones be fixed (or be able to float)? "Will the future site only have a modern vision?" All "certain" intentions can be reflected in order to better understand the site.

Three steps can be taken to better identify unique identity of the project. It is first necessary to experience a number of different appearances to get some typical forms, such as "misty and quiet" vast sea, "cloud-enveloped" Jinshan Islands, "Organic Form" of status-quo fishing village closely associated with water and dried seafood along waterfront roads, etc. These appearances indicate a common orientation. Second, it is necessary to describe the same characteristics of these appearances with one or two words so as to achieve empirical universal. For example, "sea, island, village" can be used to describe the above-mentioned appearances. Finally, efforts should be made to identify such universals to obtain some identities of the object and achieve eidetic universal. For example, elements like villagers, history, event and seawater can best reflect an understanding of the site's unique identity. Additionally, the advantages, disadvantages, potentials and challenges of the site should be made clear in this stage.

2.2.4 Preliminary Concept

Based on the previous analysis, eidetic universal can be upgraded to form a preliminary design concept. Eidetic universal, perceived or seen by different planners and designers, may be different. Therefore, design concepts can diverge widely.

In the Jinshan District (Shanghai) project, students put forward the planning and design concepts based on their perceptions, such as "Back to the sea", "Organic Urbanism", "Variation of Space—Floating Island", "Rainbow Belt" and "Three-dimensional Game". "Back to the sea" echoes the historical significance of Jinshanzui Village as Shanghai's first and last fishing village. The concept emphasizes the space-time specificity of the coastal Jinshan New Town, attempting to comprehensively integrate relations among village, history and events, and creating more coastline interfaces and waterfront landscapes (Figure 2.7). "Organic Urbanism" gives consideration to local characteristics (e.g., natural ecological texture, ocean culture characteristics and Jinshan Islands) and focuses upon waterfront urbanism based on an organic regeneration philosophy. "Variation of Space—Floating Island" focuses on the perception of spatial texture change from the city and village to sea, and creates a floating island area in the coastal reclamation zone. "Rainbow Belt", inspired by the current monotonous and thick wave parapet, serves the purpose of solving the traffic problem of the Shanghai-Hangzhou Highway. With the aim for both a versatile and complex-space, this place will become a unique landmark. The concept of the "Three-dimensional Game" attempts to construct different platforms through a dialogue with the waves of the sea and a view to the islands.

2.2.5 Conceptual Analysis

Upon access to a preliminary design concept, there is a need to further elaborate on the design concept, and deepen and enrich the connotations of the design concept. After giving full consideration to macroscopic, mesoscopic and microscopic analyses in the initial phase, designers should combine the design concept, site issues and main functions together, and infer the main design strategies which will play a guiding role in the generation of the spatial form.

After the design concept of "Back to the sea" was proposed, this group of students analyzed the attitudes and functional demands from different people, and presented different space design strategies according to the connotations of the concept (Figure 2.8): (1) Making "Strip-like" waterfront landscape associated with the locality. This strategy would help integrate expansion of the coastal areas, resume estuary ecological balance, and develop new sustainable urban coastal-style landscape;

图 2.7 规划设计概念"回归海洋"对人、时间、事件、物因素的诗性表达

图片来源：课程教学学生绘制

Figure 2.7 Poetic Expression for Man, Time, Event and Thing in Planning and Design Concept entitled "Back to the sea"

Source: students' drawing

胡塞尔发展了两种还原之路，即存在论的还原之路和笛卡尔式的还原之路。存在论的还原之路是从科学探索之中转入一种新的反思性的姿态，思考科学探索之中已经运用却没有使之成为论题（proposals）的各种意向性。比如针对金山基地，可以就对象的本体性不断地追问：生活岸线意味着什么？"城市"生活岸线意味着什么？南面的大海（对渔民和游客、对金山、对上海）意味着什么？（具有古老的历史）渔村意味着什么？要回答这些问题，需要沉思、关联和想象，由此可以演绎出各种新的认知和理解，构成最终设计概念的基础。笛卡尔式的还原之路是采取"尝试去怀疑"各种意向，而非笛卡尔式的普遍怀疑的态度。比如针对填海区的一般做法可以质疑：填海区就是用土将海面全部填塞吗？古老的渔村必定会被高容积率的商业性开发替代吗？填海区内的联系一定是陆路吗（利用水道呢）？填海区内的建筑一定都是固定的吗（可以漂浮吗）？基地的未来愿景一定是现代性的吗？诸如此类，对所有"确定性"的意向都可以进行反思，由此深刻地解读基地。

在经过上述还原之后，可以将意向从三个层面演进，以获得对象的特质。首先，连续经验（experience）一些不同的显像，获得一种典型性显像，如"迷蒙宁静"的辽阔海面、"烟云笼罩"的金山三岛、与水关联密切的"有机形态"的现状渔村、临海公路上的海鲜海产品干货等等，这些显像都涉及一种共同指向；其次，用某个词语描述这些显像的相同特征，达到一种经验的共相（empirical universal），如可用"海、岛、村"等词语描述前面的诸多显像；最后，努力辨识此类共相，获得对象的某种特质，达到本质共相（eidetic universal），如可以用村民、历史、事件、海水等因素浓缩对于基地特质的认知。

此外，这个阶段还应该整理基地的优势、劣势、潜力和挑战。

2.2.4 初步概念

在前期分析的基础上，将本质共相提升，可以形成初步的设计概念。不同的规划设计人员感知或明见到的本质共相可能不同，因而其设计概念也可大相径庭。

在上海金山区项目中，同学们按课程教学要求，提出了基于自我感知的规划设计概念，比较突出的有"回归海洋"、"有机都市主义"、"空间的变异—浮游的岛屿"、"彩虹带"、"立体游戏"等。"回归海洋"紧扣金山嘴村落作为上海第一个也是最后一个渔村的历史意义，强调金山新城滨海的时空特殊性，拟将村民、历史、事件、物因素综合考虑，将海岸线引入腹地，创造出更多海岸线界面和滨水景观（图2.7）。"有机都市主义"从金山嘴村落的自然生态肌理、海洋文化特色和金山三岛等地方性特征出发，充分依托现有村落人气，构筑一种基于有机再生理念的滨水都市主义，意图采用外部围合、内部岛屿的填海方式，强调历史性和地域性，赋予基地独特的亲水性和社区感。"空间的变异—浮游的岛屿"运用形态学原理，通过感知从城市至村落至大海的空间肌理变化及心理意象，在滨海填海处拟呈现为漂浮的岛屿形态。"彩虹带"从现状单调厚实的防浪墙得到启发，结合解决沪杭公路过境交通的问题，拟打造一条多功能、空间复杂多变的'彩虹'带，成为独特性的地标。"立体游戏"是基于基地现状的不同标高，规划设置不同的地面及空中平台，与南部大海三岛形成微妙的对话关系。

2.2.5 概念解析

在获得初步的设计概念后，需要进一步清晰地阐明设计概念，深化和丰富设计概念内涵，综合前期的宏观、中观和微观分析，将设计概念与基地问题和主体功能结合起来，推衍出主要的设计策略，以指导下一步的空间形态生成。

在提出设计概念"回归海洋"后，该组同学进而分析不同人群对于海洋的态度和功能需求，依据概念内涵，提出不同的空间设计策略（图2.8）：（1）与地方性关联的"条状"

(2) Strengthening spatial relationship of integration between "water" and "city". The aims of this strategy was to open sightseeing corridors between hinterland and sea, and remove stairs and vertical dykes blocking line of sight on the basis of solving the flood control problem; (3) Restoring local and public characteristics of "waterfront space", particularly through establishing a fishing port in order to continue the historic lifestyle of a fishing village and serve the purpose of tourism. Every design strategy can derive a corresponding morphological relationship. By means of overlapping and integrating these different relations layer by layer, overall spatial morphological structure and master plan of planning and design can be logically generated for the site.

The basic generation framework of urban planning and design concepts can be summarized with the five steps among which field research, eidetic universal, and preliminary concept are the most significant parts. Field Research concerns how the people on a site perceive the space in its present status. Eidetic Universal means to make a more in-depth understanding in their perception of things. Preliminary Concept upgrades "Eidetic Universal" into a design concept. Because these three steps are closely-related, sometimes they can be combined. Also, people can repeatedly enter a space and reflect upon it until they reach clarity of the perceived object. It should be noted that traits and attributes of spaces are quite profound and we are unable to reach an exhaustive understanding of an object. Whichever appearances this object shows us, there are always other aspects hidden in plain sight. We are not only constantly surprised to see some characteristics of the site, but also are constantly surprised to see "other aspects" presented to us in different ways. Therefore, for any site there are a variety of design concepts and solutions in the planning and design process.

图 2.8　规划设计概念"回归海洋"的设计策略
左边为现状，右边为规划设计意象
图片来源：课程教学学生绘制

Figure 2.8　Design Strategy for Planning and Design Concept of "Back to the sea"
Left: *Status quo* ; Right: Image for planning and design
Source: students' drawing

注：上图，填海的区域将由一条条伸入海中与水系和地方性关联的滨水景观构成；中图，解决防洪问题的基础上，消除阻碍视线的台阶和垂直堤坝；下图，将临水的过境道路改造具有地方性特质的渔港，既能作为生产用途，也可服务于旅游之用。

Note: As shown in the top pictures, the reclamation zone will consist of waterfront landscapes extending into the sea and link water systems and local areas. As shown in the middle pictures, stairs and vertical dykes blocking lines of sight will be removed on the basis of solving the flood control problem. The lower pictures show waterfront through roads rebuilt into a fishing port with local qualities, which can both serve the purpose of production, but also contribute to tourism.

References:
1. Dovey, K. D. "Putting Geometry in its Place: Toward a Phenomenology of the Design Process," in D. Seamon ed., Dwelling, seeing and Design: Toward A Phenomenological Ecology. Albany: State University of New York Press, 1993.
2. Holl, S. Anchoring. New York: Princeton Architectural Press, 1989.
3. Holl, S. J. Pallasmaa and A. Perez-Gomez. Question of Perception—Phenomenology of Architecture. Architecture and Urbanism [J]m, July, 1994, Special Issue Schulz. Norberg-C. Genius Loci: Toward A Phenomenology of Architecture. New York: Rizzoli, 1980.
4. Tong, Ming. Architectural Structure and Its Questions: Interpretation from Perspective of Phenomenological Analysis// Peng Nu, Zhi, Wenjun and Dai, Chun (Eds.). Dialogue between Phenomenology and Architecture. Shanghai: Tongji University Press, 2009.

滨水景观构成主要的填海方式，将滨海地区的扩建、河口生态平衡的修复、海上防御的改善和城市开发结合起来，发展新的可持续的城市海岸风情景观；（2）强化"水""城"一体的空间关系，打通腹地与海面的视线走廊，在解决防洪问题的基础上，消除阻碍视线的台阶和垂直堤坝；（3）恢复"临水空间"的公共性和地方性特质，设置渔港，延续渔村的历史生活形态，同时也可服务于旅游之用等。每条设计策略可以衍化出相应的形态操作关系，将这些不同形态关系层层叠加整合，就可以较为逻辑性地生成基地的总体空间形态结构和规划设计总平面。

城市规划设计概念基本生成框架的五个步骤中，"现场调研→本质共相→初步概念"这三步最为关键。现场调研是在场状态下，在场者通过感知，对事物进行本质直观；本质共相是对事物的感知进行更为深入的理解，通过存在论式和笛卡尔式两种还原以及典型性显像、经验共相和"本质共相"三个层次的意向演进，获得对象的特质（或潜质）；初步概念是将"本质共相"提升为设计概念。由于这三个步骤关联密切，有时也可以合并，或者多次进入现场，反复思索，直至到达所知觉对象的"无蔽状态"，获得澄明。需要指出的是，事物的特质是有深度的，我们无法穷尽一个对象的认识，无论它已经向我们呈现了怎样的显像，总还是存在其他未曾呈现的方面。我们不断因为看到某个事物的某些特征而惊讶，也因为看到它能够是别的样子、能够呈现给我们它的"其他方面"而感到惊奇。因此，对于同一个基地、同样的问题，在规划设计上可以有多种设计概念和解答方案。此外，还有其他设计方法生成城市规划与设计概念，由于篇幅所限，这里不再赘述。

2.3 Urban Planning and Design for International Students: Production Requirements for Design Outcomes
2.3 面向国际留学生的城市规划与设计：设计成果制作要求

陈竞姝　Jingshu Chen

In this course, production and expression of design outcomes are the biggest touchstones demonstrating both the teachers' ability and the students' expression. They are also a wonderful demonstration for the entire teaching process. The final outcomes are reviewed by a professional jury composed of course teachers, other invited faculty members, designers from design companies, clients and others of various backgrounds. This not only enables students to listen to different views, but also creates a professional atmosphere with greater practical significance. At the same time, the process-style teaching method designed for this course is emphasized to students at the beginning. Not only are the final outcomes evaluated, but so are the independent and group presentations in every round and class performance. Final outcomes turn into a means for student groups to embody their individuality, creativity and teamwork ability rather than merely a means to hand in "homework". The competition-centered open evaluation mode allows students to spontaneously plunge into the design, learn independently, and truly take the initiative in this course.

In the requirements for design outcomes, course setting pays equal attention to both uniformity and flexibility. The outcomes require unified forms to ensure the fairness and integrity for evaluating student groups. Moreover, in view of international students' active thinking and innovative competence, no restraints are imposed on ways of expression so as to endow student groups with space for innovation.

2.3.1 Goal of Design Production

The main goal of design outcomes is to guide students to make use of diverse expression methods to render complete and adequate expression for design philosophy, and to encourage students to try innovative approaches and learn new skills. In this way, students and student groups make their own semester summary. Through review and comment by the professional jury, students can reflect upon the entire learning process.

2.3.2 The Content of Design Production

The requirements of the final design outcomes contain three parts: display board, group presentation (PowerPoint) and booklet. It takes approximately 2-3 weeks to produce final design outcomes. A benefit to this short time period is it causes students to be wholeheartedly dedicated to the tasks and helps them fully understand the importance of teamwork.

2.3.2.1 Expression Focus of Design Production

(1) Site Study

The following topics should generally be covered:

a. Site Location at macroscopic urban level;

b. Traffic conditions, ecosystem, spatial structure and other aspects at mesoscopic regional level;

c. Natural conditions, construction status, cultural status and other existing aspects at microscopic site level;

d. Site-related cultural features, historical context, etc;

e. Site-related status quo population characteristics.

At the final outcome expression stage, the site analysis drawing is not subjected to a fixed format, but must articulate design methods and design processes.

(2) Design Concept

After individual design concept generation and groups design concepts generation at the second phase of the course (Design Concept Generation), students should submit a final design concept diagrams by teams. A "design concept" mainly describes the development process of design ideas and programs, which can be expressed in a variety of forms. This requires student groups to put forward definite concepts and titles, and elaborate on concepts so as to make them easily understood. Concept titles can be long or short (a phrase is preferred) in both Chinese and English.

(3) Analysis of Function and Structure

The analysis diagram of function and structure probes into the students' ability to synergize design concept and spatial layout. Students are required to divide functional partitions according to functions of buildings and lands, explain the relationship between various types of partitions, indicate functional axis, visual corridor, public space, key node, path, landmarks and other factors, and describe impacts of these key elements on the development of design programs.

(4) Master Plan

Students are required to submit a colorized master plan at around a scale of 1: 2000. The main elements illustrated by the master plan include: building cluster form (building outline, the number of storeys, altitude relation, etc), traffic lines and boundaries (vehicle roads, pedestrian roads, railways, highways, etc), natural environment (water body, topography and geomorphology), public space (plazas, green spaces, sports venues, etc), as well as compass (or wind rose), measuring scale, required text description, economic and technical indicators, etc.

设计成果的制作和表达是整个课程中最考验教师控制能力和学生表达能力的环节，也是对整个教学过程的精彩展示。尤其最终成果将由课程教师、校内邀请教师、设计公司设计师、甲方等不同背景人员组成的专业评审团进行点评，这不仅使学生能听取多方面的意见，也营造出一种更具有实践意义的专业氛围。与此同时，在课程之初就向学生强调本课程设计的过程式教学方法，不仅仅以最终成果进行评判，每一轮的独立汇报和小组阶段成果汇报以及课上表现都将作为考核的重要内容。因此，最终成果更多的成为学生小组展现其个性、创意和协作能力的载体，而并非简单的"交作业"。充满竞争性的公开评判方式使学生们自发地投入到设计工作和自主学习之中，真正成为课程中的主角。

在制定设计成果的要求上，课程设定兼顾统一性和灵活性。以统一的形式要求保证学生小组考察的公平性和完整性；同时针对国际学生思路活跃、创新力强的特点，以不限定表达内容和重点的方式充分给予学生小组创新和自由展示的空间。

2.3.1 设计成果的目标

设计成果制作的主要目标是引导学生使用多元化的表达方法对设计理念进行完整和充分的表达，并鼓励学生进行各类创新方法的尝试和新技能的学习。同时也是学生和学生小组的学期总结，并通过专家评审团的评审和点评，使学生们对整个学习过程有所反思和回溯。

2.3.2 设计成果的内容

最终设计成果要求分为三项：展板、汇报 PPT 以及文本。其成果制作的时间约为 2-3 周，在较短的时间周期完成三项工作并且进行最终汇报准备，使得学生有较强的紧迫感，激发其全心投入并且充分感到团队工作的重要性。

2.3.2.1 成果的重点表达内容

（1）基地分析

一般应阐述以下内容：

a. 宏观城市层面的基地区位条件；

b. 中观区域层面的交通条件、生态系统、空间结构等；

c. 微观基地层面的自然条件、现状建设状况、人文状况等；

d. 基地相关的文化特色、历史文脉等；

e. 基地的现状人群特征等。

在最终成果表达阶段的基地分析图纸并无固定格式要求，能表达清楚研究方法和研究过程即可。

（2）设计概念

设计概念部分是对学生在课程第二阶段——设计概念生成阶段中经过个人方案构思，以及小组概念生成两个环节以后，最终以小组为单位提交的设计概念图。设计概念主要说明设计思路、方案形成的过程，可采用多样的表达形式。要求学生小组必须提出明确的概念和标题，并且针对概念进行阐述，便于观者理解；概念标题可长可短，但尽量控制为一个词组，采用中英文表达。

（3）功能结构分析

功能结构分析图考察的是学生将设计概念与空间布局协同的能力。要求根据建筑物、用地的使用功能划分功能分区，说明各类分区之间的关系以及反映功能轴线、视觉廊道、公共空间核心、重要节点、路径、地标等要素，并说明这些重要元素对方案形成的影响。

(5)Three-dimensional Rendering

As an important component of design outcome, three-dimensional space rendering covers the overall bird's eye view 3D rendering (perspective view or axonometric view), and regular rendering etc. Diversification is encouraged for specific forms of expression. Students can take advantage of 3DMAX, Sketch-up and other three-dimensional computer graphics tools, as well as freehand sketching, collage or other forms of expression.

(6) Road Traffic Planning

Road traffic planning mainly covers the following content:

a. Mode of connection between internal and external traffic of the site must be made clear.

b. Design of site road system and traffic organization, analysis of road grades, different traffic flow lines (motor vehicles, non-motor vehicles, pedestrian lines, two-storey corridors, waterborne tourism lines, etc), and public transport lines and stations.

c. Parking spaces for motor vehicles and non-motor vehicles.

(7) Public Space Planning

Focused research should be made on streets, plazas, pedestrian streets, public green spaces, rivers and public external spaces between buildings (Zheng, 1995), and mainly covers the following content:

a.Cultural activity, content, scale and other aspects of public space should be specified. Spatial layout should be made in two-dimensionand three- dimension and three-dimensional perspective drawing for the key area (node) should be provided.

b.Landscape system, space scale control, interface design guide, waterfront environment and other aspects of public space should be considered, which can be freely expressed according to design concept.

c.According to the area of the site and design depth requirements, circulation, entrances, building facade, street profile, featured landscape, plant furnishing and other aspects of public space can be selectively expressed.

(8)Public Facility Planning

Major public facilities of the site should be arranged, including education facilities (nurseries, kindergartens, schools and training institutions), healthcare facilities (hospitals, clinics, etc), cultural facilities (theaters, libraries, concert halls, cultural activity centers, etc), sports facilities, social welfare facilities, commercial facilities, etc.

In addition to the above drawings, students are allowed to freely provide other drawings, diagrams or charts according to their design concepts.

2.3.2.2 Format and Requirements for Design Production

(1)Display boards

Four or five A1-size color printing boards are required. Before the final presentation, these boards should be posted on a display wall in the classroom according to sequence. The professional jury can evaluate and analyze the design outcomes of student groups before and after their presentation. These displays can also promote mutual observation and study between student groups. Display content of the boards should be dominated by drawings, coupled with brief text description and clear markings.

(2)Group presentation (PowerPoint)

Compared with boards, presentation PowerPoint provides more freedom for students. The requirements for PowerPoint contents are basically unconstrained. However, it is recommended that students should arrange the content in the following order:

Macroscopic, mesoscopic and microscopic analysis for the site—Perception of the existing situation of the Site—Generation of design concepts—spatial implementation of design concept and functional structure—Master Planning layout—Expression of design Features—Design summary.

Throughout the entire design course, the PowerPoint presentations are carried out for many rounds, and opinions are solicited from instructing teachers, clients and other experts. During this process, student groups also constantly adjust and improve their outcomes. At the final stage, teachers should primarily guide student groups to summarize key points and show highlights of the design. Student groups need to carefully take into account how to express their design outcomes in a clear and concise manner. This not only requires students to have good verbal skills, but also requires a well-organized, logical and graphic expression of the PowerPoint presentation.

(3)Booklet

A4-size design booklet is also required for the final production. Booklet should not only include the main drawing content included in aforementioned boards, but also should cover outcomes of all stages. Compared with the relatively limited space of the boards and PowerPoint presentations within a limited time period, the booklet can bring together students' more widespread design thoughts and also records the students' entire learning process. When a student group makes their PowerPoint presentation, they should simultaneously submit the booklet to the jury for review which will offer a better understanding

（4）总平面图

要求必须提供一张1：2000以上的总平面彩图。总平面图的主要图示内容包括：建筑群体形态（建筑轮廓、层数、高低关系等），交通线路和边界（车行道路、步行道路、铁路、公路等），自然环境（水体、地形、地貌），公共空间（广场、绿地、运动场地等）以及指北针或风玫瑰，比例尺，必要的文字说明和经济技术指标等。

（5）三维空间效果图

三维空间效果表达作为设计成果的重要组成，包含总体鸟瞰三维效果（透视图或轴测图）、局部透视三维效果以及人视点透视效果等。具体表达形式鼓励多样化。学生可采用3DMAX、SketchUp等计算机三维制图，也可采用手绘、拼贴或者其他表现形式。

（6）道路交通规划图

道路交通规划部分主要阐述以下内容：

a. 明确基地内部交通和外部交通的衔接方式；

b. 设计基地内道路系统和交通组织，可按道路等级、不同交通流线（机动车、非机动车、步行、二层连廊、水上游线等）、公共交通线路和站点等进行分解分析；

c. 根据需求配置机动车、非机动车停车场地。

（7）公共空间设计

对包括街道、广场、步行街、公共绿地、河流以及建筑之间的公共外部空间进行重点研究（郑正，1995），主要阐述以下内容：

a. 明确公共空间的人文活动性质、内容、规模等，从平面和三维进行空间布局，提供节点平面设计图和三维透视图；

b. 考虑公共空间的景观组织、尺度控制、界面处理、滨水环境等内容，可根据设计概念进行自由的表达；

c. 根据基地面积规模和设计深度要求，对公共空间的流线、出入口、建筑界面、沿街轮廓、景观小品、植物配置等可以有选择地进行表达。

（8）公共设施规划图

对基地内主要的公共设施进行布局，包括：教育设施（托儿所、幼儿园、中小学校、培训机构），医疗设施（医院、诊所等）、文化设施（影剧院、图书馆、音乐厅、文化活动中心等）、体育设施、社会福利设施以及商业设施等。除了以上所述的主要图纸内容外，学生小组可结合设计概念进行自由表现。

2.3.2.2 成果的主要形式和制作要求

（1）展板

展板要求制作4-5张A1规格的彩色打印图纸，在最终成果汇报前，按照展板顺序张贴于汇报教室的展示墙上，一方面可方便评审团在听取汇报前后对各组成果进行评析，另一方面也可促进学生小组之间相互观摩和学习。展板上展示内容要求以图纸为主，结合文字说明和清晰的图纸标注。

（2）汇报PPT

汇报PPT的制作和展板相比，有着更大的自由性，其制作内容要求基本没有限定。但建议学生按以下秩序进行组织：从宏观、中观到微观的基地分析——基地现状的认知——设计概念的生成——设计概念的空间落实和功能结构——设计总平面图——设计特点表达——设计总结。

整个课程设计阶段经过多轮的PPT形式汇报，听取了包括授课教师和其他邀请教师、甲方等方面的意见，学生小组不断地调整和改进自己的成果。在最终成果阶段，教师主要引导学生小组进行重点归纳和亮点展示。学生小组需要仔细考虑如何简短清晰地表达自己的设计成果。这不仅需要学生具备良好的口头表达能力，也同时要求汇报PPT的制作有着良好的组织性、逻辑性和图示表现力。

（3）文本

文本成果制作要求为A4尺寸左右的设计图册，内容除需要包括前述展板中的主要图纸内容，也可将整个课程设计过程的各个阶段成果都进行表述。相对展板有限的篇幅和汇报PPT限定时间的重点呈现，文本可以容纳学生们更加丰富的设计思考，作为展板和汇报PPT的补充，同时也是学生整个学习过程的记录。设计小组在汇报PPT的同时将文本递交给评审团进行传阅，以便更好地了解和理解其设计思路和工作历程。

of the design ideas and work processes.

2.3.3 Expression Modes of Design Production

Expression of design outcomes is not only a process of making design concepts and thinking processes concrete and tangible through the use of various drawing tools, but also the foundation for designers and related personnel to exchange opinions. Under normal circumstances, expression of design outcomes includes design representation and dissemination. Design representation includes graph, drawing, model, table and text description, while design dissemination is mainly made by communicating verbally and dubbing the multimedia presentation. By virtue of rational analysis capability and emotional expression capability, student groups can reasonably select expression modes, and effectively convey design concepts, thinking processes and technical capabilities (Figure 2.9).

Figure2.9 A Variety of Ways to Express Design

(1) Expression with hand drawing:

For hand drawing, students must grasp fine arts skills and have the ability to use sketch, pen drawing, colored pencil, watercolor, gouache, marker pen, oil pastel and other tools and techniques for the purpose of expression. Generally, these tools and techniques can be applied in design sketches, design concept drawings, perspective effect drawings and other drawings, and may also be extended to the master plan and other analytical drawings(figure2.10).

(2) Expression with computer drawing

For computer drawing, students must master the appropriate computer-aided design skills, including graphic production, image processing, word processing and layout design. The primary drawing and editing software includes CAD, Photoshop, CorelDraw, Illustrator, In Design, etc.

(3) Expression with computer model

Computer-aided modeling software is applied to make three-dimensional models. Currently, urban design professionals often use 3D Max, Sketchup, Maya and other software programs. However, new software is constantly being developed and put into use for 3D model creation. Production of computer models demand students' have a high level of software application skills. Although this course is not aimed to impart specific drawing skills, mutual learning between students with different expertise can be achieved through assignment of responsibility in the teamwork.

(4) Expression with crafted model

Depending on the specific requirements for design issues, it is advised to selectively make crafted models. Production ratio of a crafted model can vary according to the size of the site (generally 1: 1000 or 1: 2000). The bottom plate could be about A1-A0 size. Model material can be determined by students at their own discretion so as to encourage diversification of expression modes.

(5)Collage image

Collage image, characterized by a combination of real photos and design ideas, is a visualized expression mode. By means of a variety of elements collated on an image, this would reflect a designer's thinking process and imagination. For example, in the design project of Tinglin Town, Jinshan District, Shanghai, students collated status quo real photos and designed leisure venues on the bridge in order to express the idea of optimizing walking space and increasing leisureVenues (Figure 2.11).

(6)Verbal expression

Verbal expression is an important way for designers and related personnel to exchange views on design outcomes. Through on-site lectures, supportive PowerPoint presentations and reporting, a student groups' language skills can be examined. This requires that student groups assign responsibilities and arrange for report procedures prior to the outcome presentation.

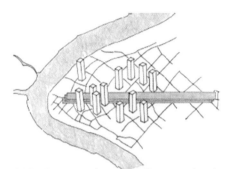

Eco-Core | Establish green spine along Century Avenue to act as catalyst for Ecocity growth

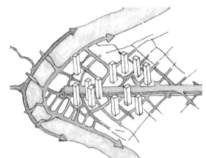

Restoration | Re-establish natural ecosystem of riverfront which will connect to the region

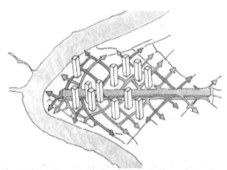

Penetration | Eco-Core moves throughout site through street network

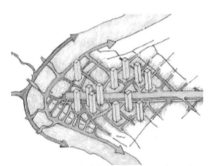

Infiltration | Blocks and buildings will be transformed to meet eco-zoning regulations established by EAR

图 2.10 手绘表达设计思路
图片来源：课程教学学生绘制
Figure 2.10 Hand-painted Expression of Design Ideas
Source: students' drawing

2.3.3 设计成果的表达方式

设计成果的表达是通过运用各种绘图手段辅助将设计概念和思维过程具象化，有形化的过程，同时也是设计师和相关人员进行交流的基础。通常情况下，设计成果的表达包含设计表现和设计传达两个部分。设计表现以图形、图纸、模型、表格、文字说明等为主要形式，设计传达则以语言沟通和配音多媒体展示为主要形式。学生小组可通过理性的分析能力和感性的表现能力，合理选择表达方式，有效地传达出设计概念、思维过程和技术能力（图2.9）。

（1）手绘图纸表达：

手绘图要求学生具有一定美术功底，可使用包括素描、钢笔画、彩色铅笔、水彩水粉、马克笔、油画棒等多种工具和技法进行表现。一般可在设计草图、设计概念、透视效果等图纸中运用，也可以扩展至总平面及其他分析图纸（图2.10）。

（2）软件绘图表达

软件绘图需要学生掌握相应的计算机辅助设计技能，主要包括图形制作、图像处理、文字处理及排版设计等。配合本课程设计的主要平面制图及编辑软件包括CAD、Photoshop、CorelDraw、Illustrator、InDesign等。

（3）计算机模型表达

运用计算机辅助三维建模软件等制作三维模型，目前城市设计专业常用的软件包括3D Max、SketchUp、Maya等，同时也有更多新的软件被研发和投入使用。计算机模型的制作需要学生具备一定的软件使用技能，虽然本课程并不教授具体的制图技能，但通过小组工作中的分工合作，可以促进拥有不同专业技能的学生之间的相互学习。

（4）实体模型表达

根据设计课题的具体要求，可以有选择性的要求制作实体模型。实体模型的制作比例可根据设计基地的大小而变化，一般可采用1：1000或1：2000的比例，底板大小在A1-A0尺寸左右。模型材质可由学生自行决定，鼓励多元化的表达方式。

(7) Expression with multimedia

In addition to the forms of expression such as PowerPoint, student groups can also freely use diverse forms of expression, such as multimedia video or animation. For instance, in the design project of Tinglin Town, a student team showed field research and interviews via a short multimedia video. Another student group demonstrated the generation process of design concept via live drama.

2.3.4 Review of Design Production

Like evaluation at the mid-term appraisal stage, review of final outcomes could also be made by the professional jury composed of other invited faculty members engaged in design, clients, designers from design companies and other experts. The jury members should comment upon the design outcomes from various perspectives related to their areas of expertise.

Students should be grouped to make presentation, and freely express for their design programs in the form of PowerPoint. Every group should complete their presentation within 20-30 minutes. All members of student groups should take part in their presentation, with a view to promoting cooperation and assigning responsibility among group members and improving and examining students' oral skills in terms of design.

After the group presentation, the jury would raise questions or suggestions and students will have 5-10 minutes to reply. Experience has shown that this phase is usually quite exciting, and the confidence of international students and their excellent communication skills deeply impress jury members.

After the presentation and student group oral defense, the jury should provide a mark according to design concept, expression of design outcome, on-site presentation, team cooperation and other aspects, finally submitting a record as a part of the student performance evaluation. In some circumstances, awards can be granted to outstanding student groups.

Review of the design outcomes includes the following parts:

(1) Design Concepts: Clarity, Innovation and Feasibility.

(2) Design Ideas: Whether the design idea is logical; whether selection of design methods is appropriate and targeted; whether comprehensive consideration has been made to address important issues, without any fundamental error; whether design idea embodies the depth of research.

(3) Design Outcomes: Whether design outcome is complete, without significant deficiency of contents; whether design drawing has prominent advantages in terms of expressive power, richness and innovation; whether graph is clear-cut and drawing is appealing (or pleasing) to the eye.

(4) Basic Skills: Whether basic design knowledge is grasped and design outcome can be perfectly expressed; whether text description is distinct and methodical; whether students have strong ability to learn and can quickly adapt to design requirements and specification discrepancies in the different cultural and geographical context.

(5) Presentation Capability: Whether students have favorable verbal skills and expressive power for presentation; whether they can concisely describe design programs within the specified deadline and clearly answer the questions raised by the jury.

(6) Cooperation Capability: Whether group members can cooperate in a satisfying manner.

References:
Zheng Zheng. On Staged Contents and Preparation of Urban Design [J]. Urban Planning Forum, 1995 (2), 26-31.

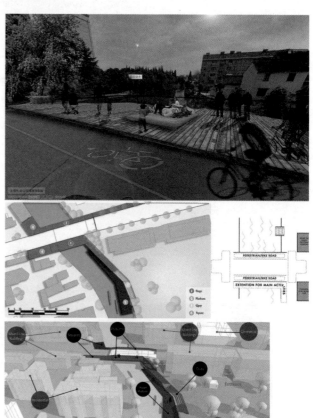

图 2.11　运用拼贴表达设计思路
图片来源：课程教学学生绘制
Figure 2.11　The Use of Collage to Express Design Ideas
Source: students' drawing

（5）意向拼贴

意向拼贴是一种结合了实景照片和设计构思的形象化表现方式，通过将多种元素拼贴在一个图面上，展现设计者的思维过程和想象图景。其表现方式非常自由，如在上海金山亭林镇的设计中，有学生将现状实景照片和设计的桥上休闲场地进行拼贴，表达其优化步行空间，增加活动休闲场地的思路（图2.11）。

（6）语言表达

语言表达是设计师和相关人员进行设计成果交流的重要途径。通过现场讲演并辅助PPT演示汇报的形式，考察学生小组的语言表达能力。需要学生设计小组在成果汇报前进行分工合作，组织好讲演流程。

（7）多媒体表达

除了PPT等表达形式外，学生小组还可自由创作多样化的表达方式，如多媒体视频或动画表达。例如，在亭林古镇的设计中，有学生小组采用简短的多媒体视频编辑展现调研和访谈过程，也有学生小组采用现场话剧表现设计概念的生成过程。

2.3.4 设计成果的评审

最终成果的评审和中期考核阶段的点评一样，也采用邀请校内其他设计教师、甲方、设计公司设计师等多方人士与课程教师一同组成专家评审团，从各种不同的角度和专业领域对成果进行点评。

学生以小组形式进行分组汇报，主要以PPT形式对各组方案进行自由表达。每组限定时间为20-30分钟。汇报要求学生小组内的每位成员都参与，促进小组成员之间的合作与分工，并且也是对学生设计口述表达能力的锻炼和考察。

在小组汇报后，专业评审团会分别提出问题或建议，进行5-10分钟的答辩环节。在历年的教学过程中，这一环节现场通常十分热烈，国际留学生的自信表达能力给参与的评审和教师们留下了深刻的印象。

专业评审团在听取各学生设计小组的汇报和进行答辩之后，依据各小组的设计概念、成果表达、现场汇报、小组合作等方面予以评分，最终提交名次记录作为学生成绩评价的一部分。在条件允许时，还可对优秀小组进行一定的奖励。

设计成果的评审要点包含以下几个方面：

（1）设计概念：明确性、创新性、可行性。

（2）设计思路：思路是否具有较强的逻辑性；设计方法的选用是否恰当和具有针对性；是否能较为全面的考虑问题，并且无原则性错误；是否具备一定的研究深度。

（3）设计成果：设计成果是否完整，内容无重大缺失；设计图纸是否在表现力、丰富性、创新性上有突出优点；制图是否清晰，图面是否整洁美观。

（4）基本能力：具备基本的设计基础知识、能较为完善的表达设计成果；文字说明是否清晰有条理；是否具有较强的学习能力，能快速回应不同文化和地域的设计要求和规范差异。

（5）汇报能力：是否具备较好的口头表达能力和汇报表现力；是否能在规定时间内简明扼要地介绍设计方案，并且清晰地回答所提出的问题。

（6）合作能力：小组成员能否进行较好地进行分工和协作。

参考文献：

郑正.论城市设计的阶段内容和编制[J].城市规划汇刊，1995（02）：26-31+45-64.

第三部分
城市设计教学的国际经验
International Experiences of Urban Design Studio Teaching

3.1 Urban Design Pedagogies in an Increasingly Globalized World
3.1 在日益全球化的世界中的城市设计教学法

Stefan Al 著　　李经纬 译　　田莉 校
Stefan Al　　Translated by Jingwei Li　　Proofread by Li Tian

The growing number of international students in Chinese universities is a sign of China's growing importance in an increasingly global world. Chinese universities will benefit from this. International students bring perspectives and ideas from other places to the table, which is advantageous to domestic students. Moreover, when many of these international students return home after their studies, their new global perspectives will help them to contribute to their fields and leave a legacy.

But while teaching international students is really a position of privilege, it also comes with responsibilities and challenges. In the following, I will briefly explain my experiences in teaching urban design graduate students at the University of Pennsylvania, where I lead the urban design concentration, and the University of Hong Kong, where I was the director of the urban design program. I will also write from the perspective as a former international student myself, having studied at the ETSAB in Barcelona, the Bartlett in London, and UC Berkeley in the United States.

First, it can be difficult to balance material in courses that is relevant for both domestic and international students. If the study material covered is solely relevant to the country of the university, than the international students will end up loosing out, learning material that might not be useful for application back home. Therefore it is important that teachers expand their curricula to include case studies and examples from other places.

For instance, I currently teach the first core graduate class in urban design, called Fundamentals of Urban Design. About half of the students are international students, and a good amount of those are from China. Therefore the case studies I show are also from places including in Asia and other continents. This approach is typical of our entire Department's strategy. Where classes used to focus more on planning in the United States, we now make a deliberate attempt to cover a wide diversity of places globally.

Moreover, it is important to realize that there is not a single country that has a monopoly on the "best" urban design projects. Different countries respond well to their own specific environmental, cultural, and economic conditions. My home country, The Netherlands, is known to outfit its cities well against flooding. Latin American countries such as Brazil have pioneered new approaches upgrading of informal settlements. Hyper-dense Hong Kong leads the world in building high-density and transit-oriented developments, outfitting large building complexes with direct subway access. Singapore, because of its limited land-supply, has innovated with urban agriculture and urban water reservoirs. Many European countries have a long tradition of dealing with cultural heritage in modernizing cities. Therefore, taking an international approach to urban design is actually better, since instructors can sample from the world's best projects, rather than from one country alone.

It is also useful to teach students general concepts and techniques that can be applied to other places. For instance, William Whyte's form of participant observation, so well narrated in his book The Social Life of Small Urban Spaces, can be applied not only in New York, where the studies were undertaken, but anywhere else in the world. Similarly, Kevin Lynch's The Image of the City gives us city elements — paths, landmarks, nodes, edges, districts — which we can apply to many other places besides Boston, where he did the original studies.

In addition, the historical trajectories of cities can be compared since the same forces of production and ideas of design have affected them. For instance, each year I co-teach an online class called Designing Cities with Professors Gary Hack and Jonathan Barnett to about 30,000 students from all over the world, including in Asia, Europe, Africa and Latin America. We start the course by giving a brief overview of how cities have evolved, from pre-industrial to industrial to post-industrial and today's mega-city region. We also teach them what ideas have influenced city building, such as modernism. We then ask them to look at their own city and to try to identify traces of these ideas and modes of production. We then discuss the result of their assignments and look and the city plans, which show that indeed the development of many cities is remarkably similar.

Taking an international approach to urban design studios also makes a lot of sense since urban design practice is increasingly becoming global. These days, large firms do projects all over the world. Urban design as a field lends itself to international approaches, different from a discipline like law for instance, where legislation in one country might not apply to others. At UPenn we teach a wide variety of international urban design studios each year, in which we work on design projects in South America, Europe, Africa and Asia.

An added bonus of international studios is that the competitive difference between domestic and international students ceases to exist, since domestic students no longer have a clear advantage. This leads me to my other point. It can be difficult to manage the wide variety in backgrounds that exists between international students and domestic students. For instance, some countries have undergraduate design degrees that take five years, others three years. Universities emphasize different ways of representation and different technological tools. This can lead to big differences in skill

中国大学里国际学生数量的增加是中国在日益全球化世界中越来越重要的标志,中国的大学也将从中受益。国际学生将观点和想法从其他地方带过来,这有利于国内学生视野的拓展。此外,当这些留学生中的一些人学成后回国,他们新的全球视角将有助于他们在自己的领域中做出贡献并传承下去。

但是教授国际学生确实不同于教授国内学生,它会带来不同的责任和挑战。我将简要地介绍一下我在宾夕法尼亚大学城市设计研究生教学中的经历,在那里我主要教授城市设计以及在香港大学我作为城市设计项目主任的经历。我也会从自己曾是一名国际学生的角度来写,我在巴塞罗那建筑学院、伦敦大学巴特莱特建筑学院和美国加利福尼亚大学伯克利分校学习过。

首先,很难去平衡与国内和国际学生都相关的课程材料。如果学习材料都是与这个大学所在的国家完全相关,那么国际学生学到的知识可能回国后不适用。因此教程有必要包括其他地方的案例研究和实例。

例如,我目前在教授研究生的第一个城市设计核心课程:城市设计基础。大约有一半的学生是国际学生,并且大部分来自中国。因此,在案例上我也包括了亚洲和其他大洲。这种方法是我们整个系典型的教学策略。之前课上更多关注美国的规划,现在我们试着去尝试覆盖全球范围不同地方的规划。

此外,认识到没有一个国家的城市设计项目是独有的"最好的"这点很重要。不同国家对自己特定的环境、文化和经济条件做出的回应也不同。我的祖国荷兰,有良好的防洪体系是众所周知的。拉丁美洲国家如巴西,在更新非正式聚落方面颇有创新。超密集的香港在世界上有着领先的建设高密度和公交引导开发的经验,地铁直接接入大型建筑综合体。新加坡由于其有限的土地供应,在城市农业与城市水源水库利用上有所创新。许多欧洲国家在现代化的城市中处理文化遗产有着悠久的传统。因此,以国际化的视角去做城市设计非常必要,老师可以列举世界上最好的案例,而不是某一个国家的案例。

教给学生可以应用到其他地方的普遍概念和技术也是有用的。例如,威廉·怀特的参与观察方法在他的书《小型城市空间的社会生活》中有很好的阐述,不仅可以在作为研究对象的纽约应用,还可以在世界的其他地方应用。同样,凯文·林奇的《城市意象》展示给我们城市的元素—路径,节点,地标,边界,街道—这可以适用于除了做原始研究的波士顿外的其他许多地方。

此外,基于类似的经济和设计理念的影响,城市的历史足迹是可以相互比较的。例如,每年我同 Gary Hack 教授和 Jonathan Barnett 教授一起教网络课程,有 30000 个来自世界各地的学生,包括亚洲、欧洲、非洲和拉丁美洲。这门课是"城市设计"。我们从简要介绍城市如何发展开始,从未工业化到工业化到后工业化再到如今的巨型城市区。我们也教学生是什么思想影响了城市建设,如现代主义。然后我们请学生观察自己的城市,试图找出这些思想的痕迹和产生的方式。我们再讨论作业和城市规划的成果,可以看出事实上许多城市的发展轨迹有惊人的相似性。

在城市设计工作室中用国际化的方法也很有意义,因为城市设计实践的全球性正日益加强。目前,大公司接手世界各地的项目。城市设计领域本身就要有国际化的方法,与其他学科比如法律明显不同:一个国家的立法可能不适用于其他国家。在宾大我们每年都开设各种各样的国际城市设计工作坊,我们的设计项目分布在南美、欧洲、非洲和亚洲的很多地方。

国际工作坊的一个额外好处是国内和国际学生之间的差异不再存在,因为国内的学生不再有明显的优势。这就涉及我的另一个观点。国际和国内学生在各种各样不同背景上存在的差异是很难解决的。例如,一些国家本科设计学位需要五年,一些国家是三年。大学强调了不同方式的表现和不同的技术工具,这会导致学生之间技能的重大差异。因此,无论是在语言、设计还是技术技能方面,确保为学生设置最低的录取标准是重要的。这也有助于面试考生来评估他们对专业的理解和语言技能。虽然当教授国际设计工作坊、学习国际化的概念和工具十分重要,但也必须意

sets among students. Therefore, it is important to make sure admission criteria set the solid minimum requirements for students, whether in language, design, and technical skills. It also helps to interview candidates to assess their understanding of the field and language skills.

But while it is important to do international design studio, and to learn concepts and tools that can be applied internationally, it is vital to realize that foreign ideas not always apply to other places. Students need to understand the local context first, including the ways in which the built environment has been shaped. To make sure our international design studios are more meaningful and realistic, we always collaborate with local planning institutions, universities and experts. We also travel to these places with our students. This is often an eye opener for students, making them rethink many of their initial assumptions.

For instance, in 2014 I was one of three instructors in the Global Housing Studio at UPenn. The first half of the semester we investigated global models for social housing, including case studies in Singapore, The Netherlands, the United States, and Chile. The second half of the semester we designed affordable housing projects for the cities of São Paulo, Toronto and Beijing (figure 3.1). Beijing's Municipal Institute of City Planning and Design had given a ten-hectare site to explore new forms of affordable housing. This government agency is tasked with planning a staggering amount of 200,000 affordable housing units every year— roughly the equivalent of building the entire housing stock of a city like Dublin, year after year. Both in terms of speed and scale, it is an unprecedented housing challenge. The Beijing team quickly realized that unlike the cities of São Paulo and Toronto, the key challenge for Beijing is managing rapid population growth and urbanization. Instead of developing ideas for that particular site alone, students recognized that the provision of affordable housing at this immense scale is really a form of city building, and that it could be used as a major impetus to improve Beijing's urban form and public realm. They tested how density zoning, urban design regulations, landscape and open space typologies, architectural forms, and economic policies could contribute to not only a more equitable, but also a more efficient, pleasant, and sustainable city (figure3.2).

Finally, I am thrilled that more international students are getting admitted to Chinese universities. It is telling of this current moment of urbanization, called by many the "Asian Century," and reflective of the international job market. I hope you will continue to admit international students. At the University of Pennsylvania we are very proud of our long history of teaching international students. One of the first Chinese students to study architecture at UPenn was Liang Sicheng in 1924, considered the father of modern Chinese architecture, whose influence lasts until today.

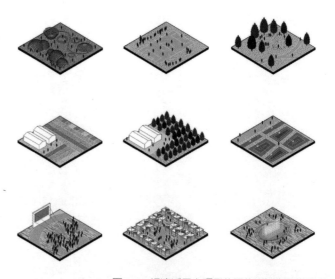

图 3.1 经济适用房项目的开放空间类型和策略
Figure 3.1 Open space typologies and strategies developed for affordable housing projects

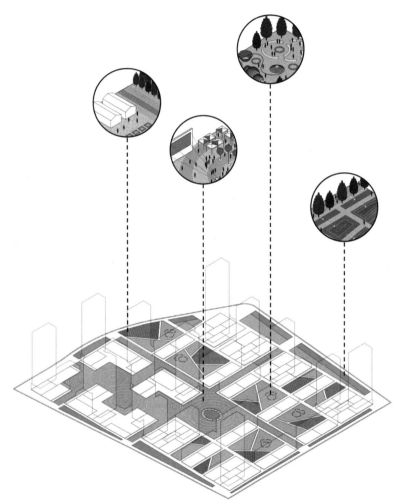

图 3.2　北京经济适用房区域开放空间类型
Figure 3.2　Open space typologies applied to the affordable housing site in Beijing

识到 A 国的设计思想不一定适用于 B 国或其他地方。学生首先需要了解当地的背景，包括建成环境是如何形成的。为了确保我们的国际设计工作坊更加有意义和实际，我们总是与当地规划机构、大学和专家进行合作。我们也同学生去这些地方旅行，这对学生来说往往大开眼界，让他们重新思考许多他们最初的假设。

例如，在 2014 年我作为宾大全球住房工作坊的三位导师之一，前半学期我们调查了全球范围内社会住房的模式，包括新加坡、荷兰、美国和智利的案例研究。后半学期我们设计了圣保罗、多伦多和北京市的经济适用房项目（图 3.1）。北京市城市规划设计研究院给了 10 公顷的土地用来探索新的经济适用房形式。这个政府机构的任务是每年规划数量惊人的 20 万栋经济适用房单元，数年的建设后相当于整个柏林市存量住房。在速度和规模上，都是住房方面的一个前所未有的挑战。北京很快就意识到，不同于圣保罗和多伦多的城市，北京面临的主要挑战是控制人口的快速增长和城市化，而不是单个的地段去提出想法，学生认识到在规模巨大的经济适用房实际上是城市建设的一种形式，并且它可以被用来作为改善北京城市形态与公共领域的主要动力。他们考察了密度分区、城市设计法规、景观和开放空间类型，了解了建筑形式和经济政策是如何有助于一个更加公平、更有效、愉快和可持续发展的城市（图 3.2）。

最后，我很激动地看到有更多的国际学生能够进入中国的大学。这也说明了在此刻许多人称之为"亚洲世纪"，城市化进程带来了国际就业市场的很多机会。我希望中国大学能继续招收国际学生。在宾夕法尼亚大学，我们很自豪我们有教授国际学生的悠久历史。1924 年梁思成是在宾州大学学习建筑的第一批中国学生之一，被誉为是中国现代建筑之父，影响至今。

3.2 Urban Design Studio at Georgia Tech: University Ave, Pittsburgh Neighborhood and the Mcdaniel Creek Watershed
3.2 佐治亚理工大学城市设计教学实例：匹兹堡社区的大学路和麦克丹尼尔河流域设计

Richard Dagenhart 著　　李经纬 译　　田莉 校
Richard Dagenhart　　Translated by Jingwei Li　　Proofread by Li Tian

This University Avenue site lies within the McDanial Creek Drainage Basin, which is the most upper part of Atlanta's South River Watershed. The site itself is vacant, having been used for major long distance trucking operations for many years. The Atlanta Beltline, a future multi-use trail and light rail transit, forms the southern boundary of the site. This site was the focus of the Beltline Subarea 2 Plan. To the north of the site is Pittsburg, a working class neighborhood that has been negatively impacted by housing abandonment, crime, and neglect. To the south of the site are several small neighborhoods.

This total area comprises the McDaniel Creek Drainage Basin. McDaniel Creek in the Pittsburg neighborhood is buried in a 6-meter diameter pipe that flows directly into McDaniel Creek south of the Beltline. When the combined sanitary and storm sewer was separated a few years ago, an unintended consequence was an increase in flooding. This occurs in Pittsburg because of the lack of stormwater capacity in the existing pipes, and flooding occurs south of the Beltline because of the volume of water flowing down the open McDaniel Creek. The site itself also floods at the lowest elevations with surface runoff from overflows from Pittsburg.

The proposal for this project is in three parts, responding to the hydrological characteristics of the McDaniel Creek Drainage Basin. First, the aim in Pittsburg is to slow stormwater, by extending the time of concentration with soft infrastructure limiting water volume entering the pipes and reducing surface runoff with infiltration. Second, the aim in the south of the site is to increase the time of concentration by moving stormwater as quickly as possible into McDaniel Creek. Hard instead of soft infrastructure would be the strategy for this area, but with very careful landscape management to accommodate rapid flows along a new McDaniel Creek Greenway and Trail. Finally, the University Avenue site would have to play an import role in stormwater detention to accommodate excess volume from Pittsburg. By partially closing the 6-meter underground pipe, the spring fed flow from McDaniel Creek can be brought to the surface to form a combined retention/detention facilty that becomes an amenity for the future development of the University Avenue site and the Atlanta Beltline.

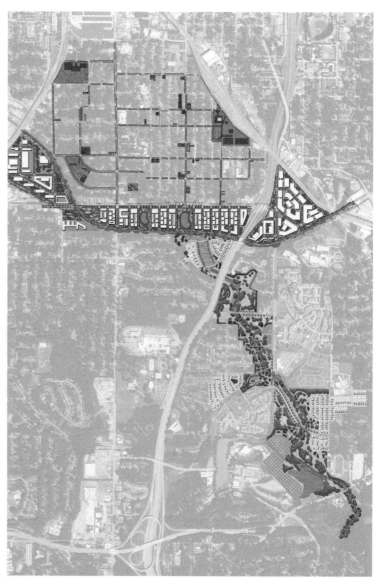

图 3.3　软基础设施总体规划
Figure 3.3　Soft Infrastructure Master Plan

大学路的基地选址在麦克丹尼尔溪流域内，位于亚特兰大南部流域最上游。它本身是空地，多年来被用于大型长途货运业务。亚特兰大的贝尔特线，一条未来的多功能步道和轻轨运输线，形成了该地点的南部边界。这个基地是贝尔特线分区2计划的重要地段。基地的北部是匹兹堡，一个被遗弃的房屋、犯罪和漠视困扰的工人阶层社区，基地的南部包括几个小型的社区。

整个区域构成了麦克丹尼尔溪流域。匹兹堡附近的麦克丹尼尔河的水流被收集在一个6米直径的管道中，这个管道的水流直接流入贝尔特线以南的麦克丹尼尔河。当几年前污水和雨水管道分离，意想不到的是造成了洪水的增加。发生在匹兹堡的原因是由于现有管道暴雨容量的不足，贝尔特线南部发生洪水时，水流冲入敞开的麦克丹尼尔河。该地段海拔最低的地方也被洪水冲袭，那是从匹兹堡溢出的地表径流。

结合麦克丹尼尔河流域的水文特征，本项目分为三部分。首先，匹兹堡地段的目的是减缓暴雨，通过"软"性基础设施，延长汇流时间，限制水量进入管道，减少地表径流的吸通性。其次，基地南部尽可能快地将暴雨水流入到麦克丹尼尔河中，减少汇流时间，这一地段的策略是用"硬"基础设施代替"软"性基础设施，但是要配合非常谨慎的景观管理，以适应沿着新麦克丹尼尔溪绿道和步道的快速水流流动。最后，大学的林荫路地段将会阻滞雨水，在调节匹兹堡过剩的水量方面发挥重要作用。关闭部分6米的地下管道，来自麦克丹尼尔河的水流可以被带到表面，形成一个合并保留/阻滞设施，为大学林荫路和亚特兰大贝尔特线未来的发展形成良好的基础。

图 3.4 基地现状
Figure3.4 Existing site conditions

图 3.5　贝尔特线区，匹兹堡社区和麦克丹尼尔河流域发展规划
Figure3.5 BeltLine Subarea 2, Pittsburgh Neighborhood Plan, and McDaniel Creek Watershed

#1 Pittsburgh Neighborhood: The problem in Pittsburgh is the stormwater in the entire neighborhood is pipes, sending a large amount of water downstream. The current proposals do not address this issue.

匹兹堡邻里：匹兹堡的问题是整个邻里的雨水管道，暴雨季节导致大量的雨水溢出。目前的方案没有涉及这个问题。

#2 University Avenue Site: The problem this site faces is the barrier it creates between the upper and lower portions of the watershed. The site is located at the point where runoff from upstream exits the pipe into McDaniel Creek. The current BeltLine proposal does not address this issue.

大学林荫路地段：这个地段的问题是上游和下游流域之间存在屏障。在该地段从管道溢出的雨水进入麦克丹尼尔河。目前的贝尔特线方案没有解决这个问题。

#3 McDaniel Creek: The City of Atlanta has several projects in the works to control flooding issues within McDaniel Creek, but the City currently has no plans to address the issues upstream that are causing the creek to flood.

麦克丹尼尔河：亚特兰大市已经着手几个工程项目，控制麦克丹尼尔河的洪水问题，但该市目前没有计划解决产生洪水的河流上游问题。

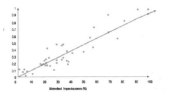 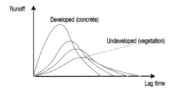 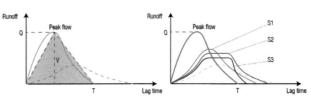

图 3.6　匹兹堡邻里软基础设施策略
Figure 3.6　Pittsburgh Neighborhood Soft Infrastructure Strategy

The coefficient relationship between runoff and imperviousness. We know that the runoff coefficient goes up when imperviousness increases. According to peak flow calculation, the higher the imperviousness, the higher the peak flow rate. When imperviousness is greater than 10%, water quality will decrease. This watershed is approximately 46% impervious.

The volumes of runoff for different impervious conditions within the watershed. The goal is to decrease the amount of impervious surfaces to minimize the runoff into McDaniel Creek.

By increasing the lag time, the peak flow is reduced and the volume of water generated during a storm reduces. Currently, the site generates 60 acre feet in a 100 year storm event, 34 acre feet in a 5 year storm event, and 12 acre feet during a 2 year storm event.

Strategy 1: Increase infiltration of water to decrease runoff by 40%.

Strategy 2: Retain water in ponds and parks to reduce runoff by 30%.

Strategy 3: Move water out of the lower portion of the watershed before the water reaches McDaniel Creek.

径流和抗渗性能系数之间的关系。我们知道，抗渗性能增加的时候径流系数明显增加。根据洪峰流量计算，越高的抗渗性能，就有越高的峰值流量。当抗渗性能大于 10% 的时候，水的质量会下降。该流域大约是 46% 的抗渗性。

该流域内不同抗渗条件下径流的水量。目标是通过减少抗透表面的数量来减少径流进入麦克丹尼尔河。

通过增加滞后时间，洪峰流量减少并且雨中产生的水量减少。目前，该地段发生百年一遇洪水时产生 60 英亩英尺的水量，五年一遇洪水暴发时产生 34 英亩英尺水量，两年一遇的洪水暴发时产生 12 英亩英尺水量。

策略 1：增加渗透水，减少 40% 的径流。

策略 2：将水贮存在池塘和公园，减少 30% 的径流。

策略 3：在水到达麦克丹尼尔河之前，将水从下游流域中排出。

Strategies for University Avenue Site

图 3.7 现有雨水管道
Figure 3.7 Daylight Existing Storm Water Pipes

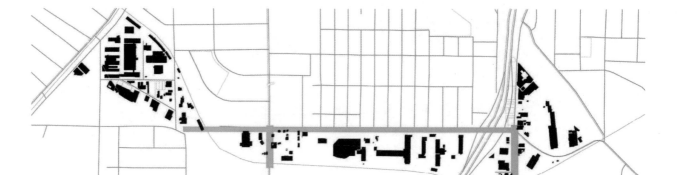

图 3.8 建设绿街道
Figure 3.8 Build Green Streets

图 3.9 贝尔特线台阶状绿带
Figure 3.9 Terrace the BeltLine for Stormwater Infrastructure

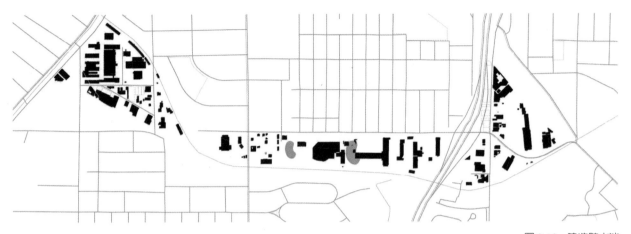

图 3.10 建造贮水池
Figure 3.10 Create Retention and Detention Ponds

图 3.11 建设新的公园
Figure 3.11 Create New Parks

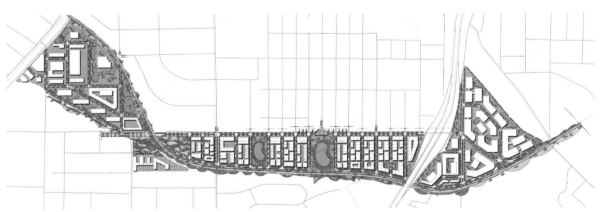

图 3.12 大学路总体规划设计
Figure 3.12 University Avenue Master Plan

第四部分
国际学生设计作业选编
Collections of Works of
International Students

4.1 上海陆家嘴中心区更新概念城市设计
URBAN DESIGN OF LUJIAZUI CENTRAL BUSINESS DISTRICT REGENERATION

指导老师： Richard Dagenhart, Perry YANG, 田莉, 李晴
INSTRUCTORS: Richard Dagenhart, Perry YANG, Li Tian, Qing Li

4.1.1 陆家嘴——房地产的游戏
LUJIAZUI : REAL ESTATE GAME

4.1.2 永续的城市——生态城市
ENDLESS CITY：ECOCITY

4.1.3 镜像城市
CITY OF MIRROR

4.1.4 公共空间与流动城市
THE PUBLIC DOMAIN & THE CITY OF MOVEMENT

4.1.5 乐活城市，乐活生活
PLAYFUL CITY PLAYFUL LIFE

上海陆家嘴基地区位
Introduction to Lujiazui Site

Lujiazui is located on the east of the Huangpu River within the Pudong district, which is the most rapidly developing place in Shanghai

陆家嘴位于上海浦东新区黄浦江东畔，是上海最具现代魅力的地方。它是上海打造国际金融贸易中心的重点建设地区。

区位
Location

Century Avenue is a boulevard going through Lujiazui and Pudong, measuring 5.5 km long and almost 100 m wide. It includes 8 lanes for automobiles and a central green corridor in the middle of the avenue, and it passes through the financial center, new city commercial center, Zhuyuan commercial district, Huamu administrative and cultura center. It aims to become the Champs Elysees in the eastern world.

世纪大道，是一条横贯浦东的景观大道。从东方明珠电视塔至浦东世纪公园，全长约5.5公里，宽100米，设8个车道和中央绿化带。它连接陆家嘴金融贸易区、新上海商业城、竹园商贸区、花木行政文化中心，为浦东新区的最重要景观大道，被誉为东方的香榭丽舍大街。

基地范围
Site Boundary

现状照片
Pictures of status quo

4.1.1

LUJIAZUI : REAL ESTATE GAME
陆家嘴——房地产的游戏

学生：Changhui ZHAO , DI Gan, Yujia QU, Ross, Julie, Renato, Edward

本设计方案的出发点是最大限度地利用公共基础设施投资，同时使个体发展潜能增至最大限度。为了达到设计目标，设计者提出了三点设计策略，分别是：减少道路宽度，建造悬臂式的建筑来增加更多的面积，通过增加建筑密度和建筑高度来增加区域密度。通过分析现状和已规划的街区数据后，我们通过增加不同的容积率提出更有效的土地利用方式。方案的终极目标是通过该方案的实施，城市能够从土地再开发中获得最大经济利润，同时建立一个更高密度并且更有活力的都市生活区。

The starting point of this design is to maximize public infrastructure investment and maximize private development potential. In order to archive the design objective, the designer put forward three design strategies, respectively being: to reduce road widths, cantilever buildings for additional square footage and increases density through more building coverage and vertical growth. The ultimate goal is to derive maximum economic profit from the land and to establish a more dense and vibrant urbanism through the scheme.

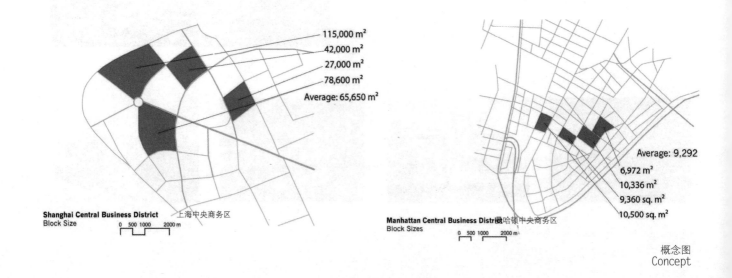

概念图
Concept

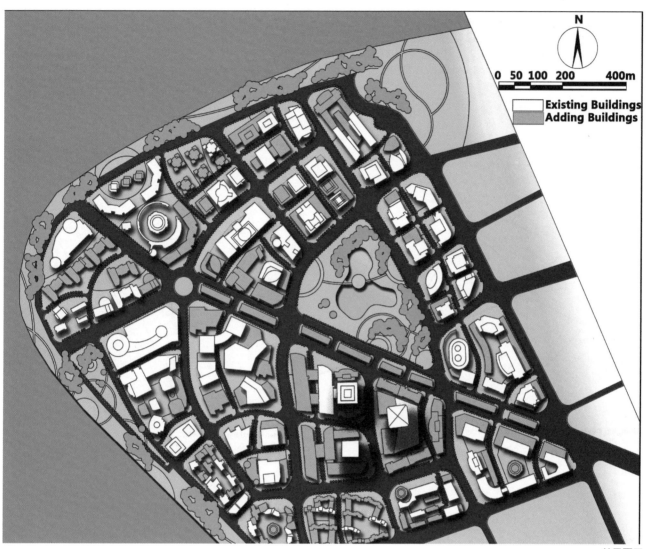

总平面图
Master plan

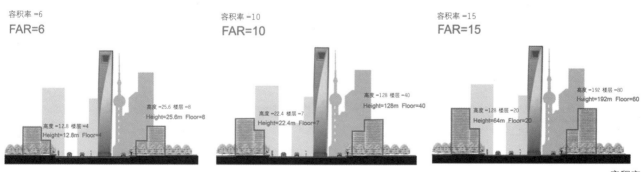

容积率
Volume Ratio

Shang Hai 上海

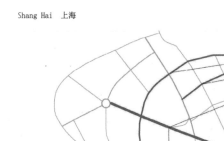

Manhattan 曼哈顿

道路宽度
Street Widths

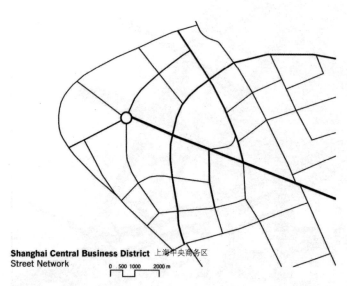

街道网络
Street Network

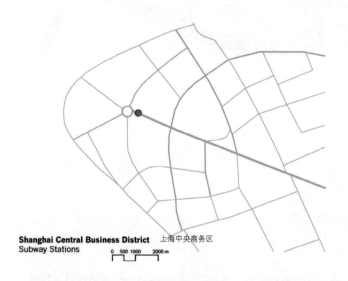

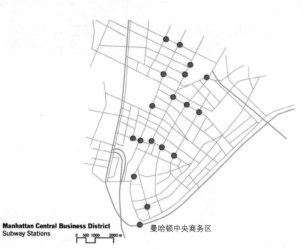

地铁站点
Subway Station

50

Strategies 设计策略

策略一：降低道路宽度，获得更多土地

Strategy 1 Reduced road width: get more lands

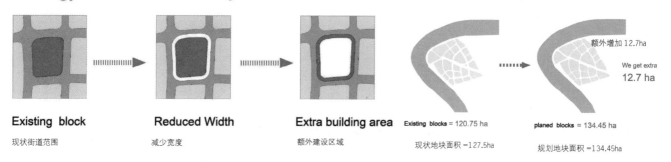

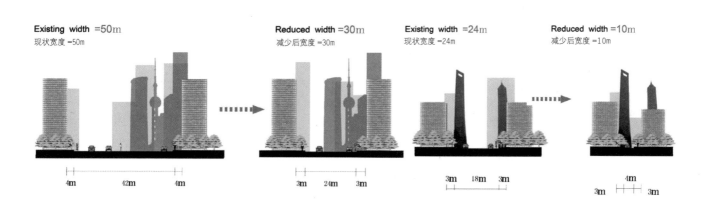

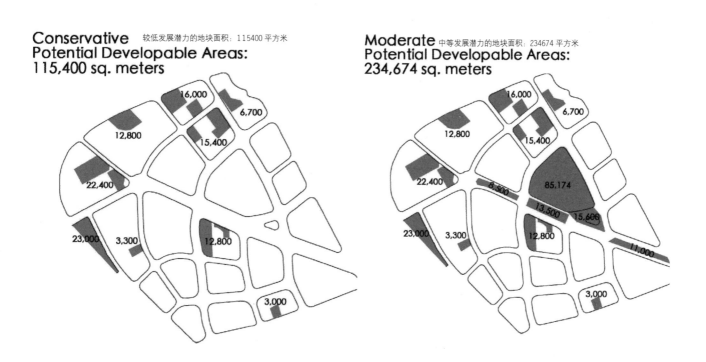

策略二：建立建筑间通道，获得更多空间
Strategy 2 Build buildings across road to get more space

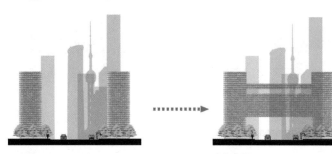

Buidlings along the road
沿街建筑

Add buidlings across the road
加入跨道路建筑

Buidlings across road to get extra areas
跨道路建筑获得额外空间

Maximum Potential Developable Area: 414,460 sq. meters

最大发展潜力的地块面积：414460 平方米

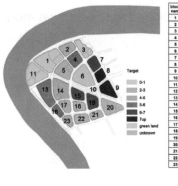

现状容积率
Existing FAR

现状建筑覆盖率
Existing Building Coverage

规划建筑覆盖率
Planned Building Coverage

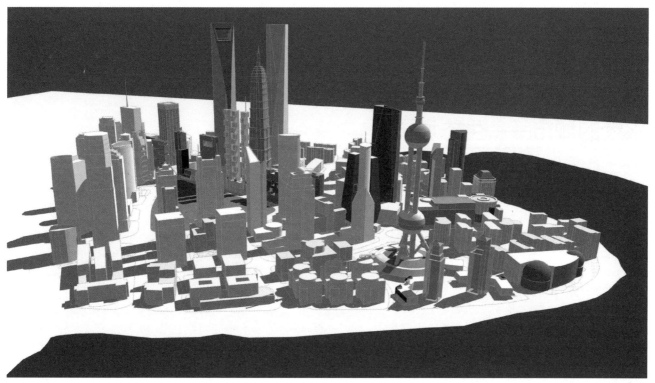

方案效果
Prospective

剖面
Section

意向1
Images One

意向2
Images Two

放大详图
Detail

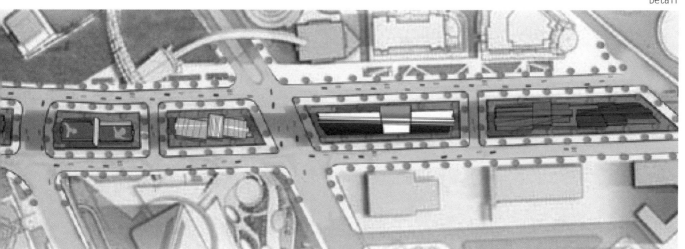

4.1.2

ENDLESS CITY: ECOCITY
永续的城市——生态城市

学生：Bo SONG, Drew Getty, Gina Mendez, Claire Thompson, Jennifer Williams, Gaylan Williams, Xiaoxue SHENG

针对当前上海社会经济发展特点，设计者意识到，城市规划本身也面临着挑战——从粗放式设计到基于研究的转型和回归。本设计方案通过逐步实现生态区、渗透、修复和融合四个基础发展策略来最终实现作者的设计意图。首先，沿世纪大道建立绿色通道作为生态城市发展的催化剂。其次，通过街道网络使生态核心渗透到整个区域。然后，重新建立沿黄浦江的生态系统并将其连接到区域。最后，规范街区和建筑的建造使其满足生态区所要求的生态容积率。

Given the transition of socio-economic development trend in Shanghai, the designers realize that urban planning itself is also facing the challenge of transformation from sprawl to compact development. The design idea is realized through a step-by-step basis of development strategy. That is Eco-Core, penetration, restoration and infiltration. First, establish green spine along Century Avenue acts as catalyst for Eco-City growth. Second, Eco-Cores are prevalent through the design of street network and natural ecosystem of riverfront is restored which connects to the wider region. Then, blocks and buildings are transformed to meet eco-zoning regulations established by Ecological Area Ratio (EAR).

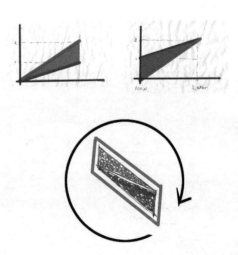

概念图
Concept

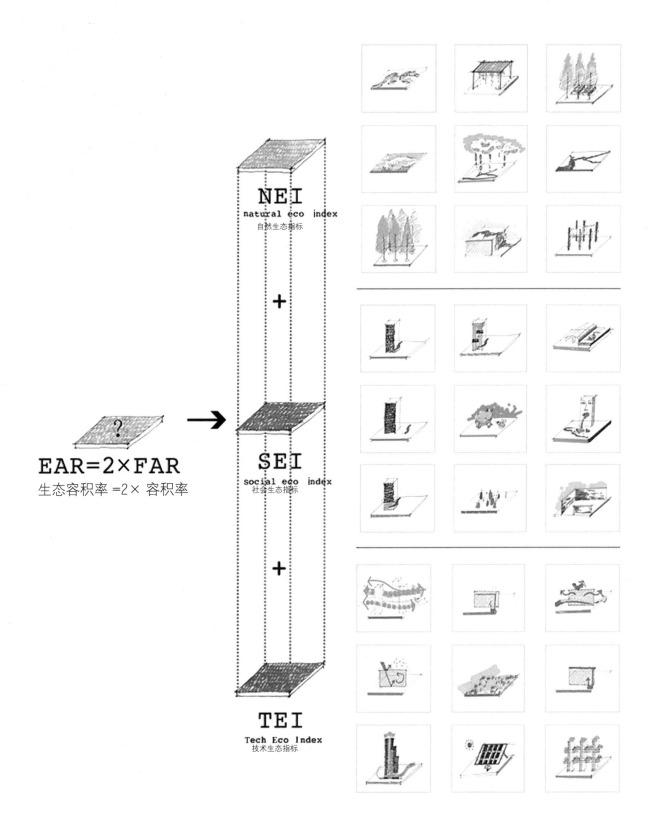

生态容积率概念
Concept

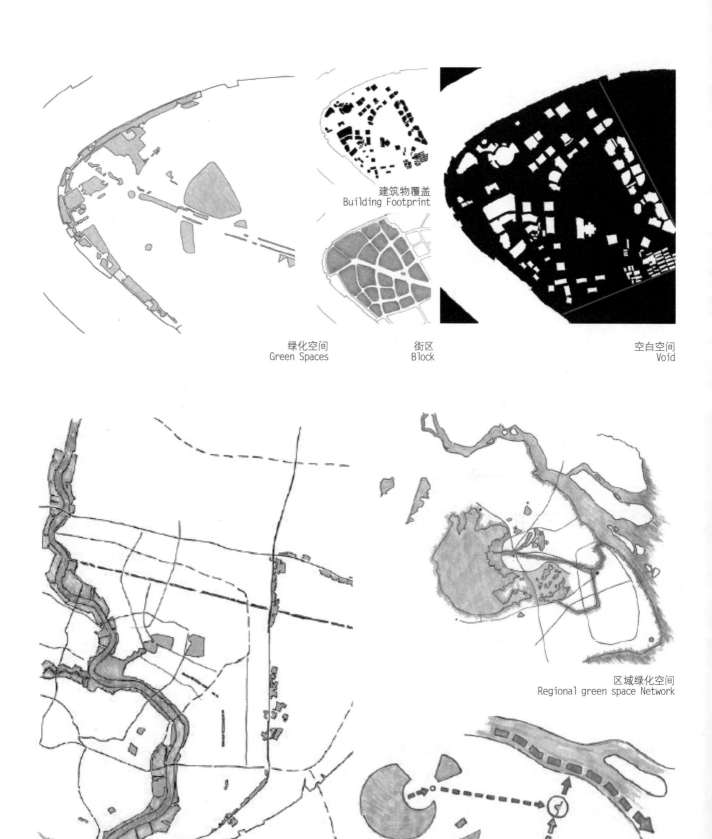

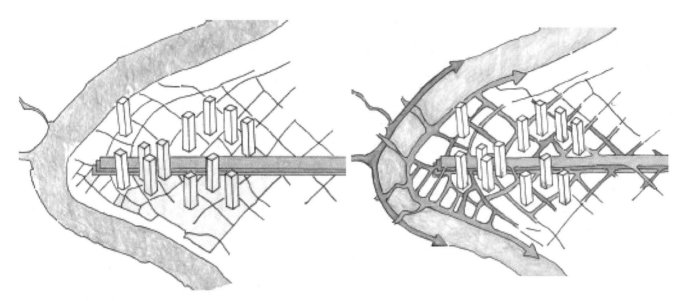

生态核：沿世纪大道设计绿轴，促进生态城市的合理扩张
Eco-core | Establish green spine along Century Avenue as catalyst for EcoCity growth

恢复：重建联系各个区域的自然生态系统和滨河空间
Restoration | Re-establish natural ecosystem of riverfront which connects to the region

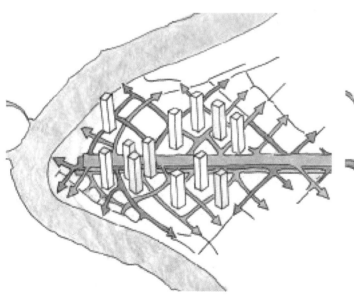
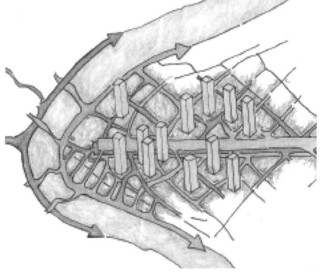

渗透：通过街道使得生态核渗透到基地各处
Penetration | Eco-cores extended through the design of network

扩散：通过改变街道和建筑的形态来满足基于 EAR（生态容积率）的控制要求
Infiltration | Blocks and buildings are transformed to meet eco-zoning regulations established by EAR

生态化改造策略
Strategy of Ecological transformation

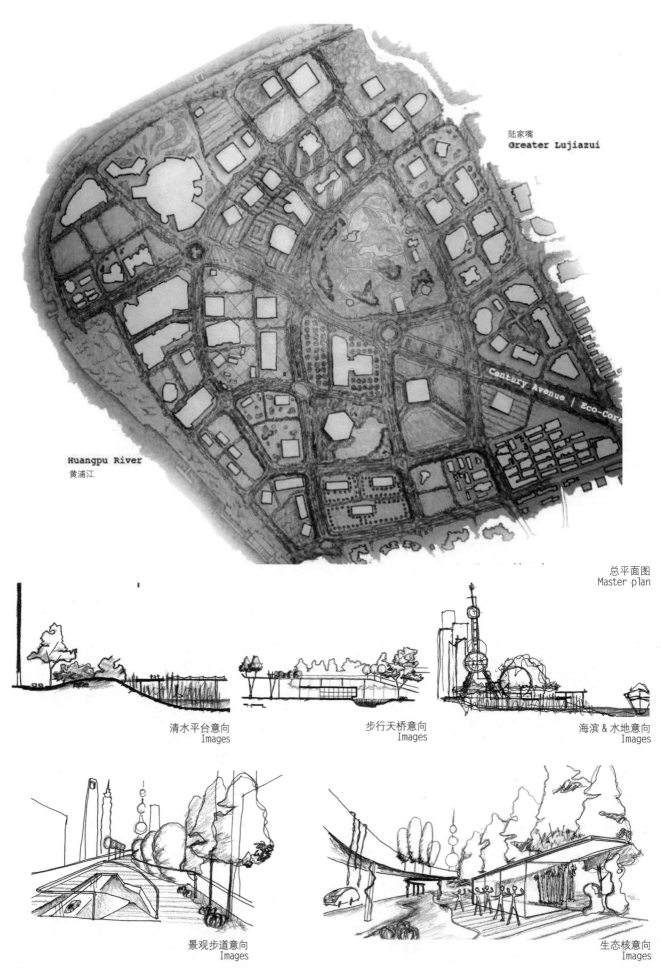

总平面图
Master plan

清水平台意向
Images

步行天桥意向
Images

海滨＆水地意向
Images

景观步道意向
Images

生态核意向
Images

通道意向
Images

滨水意向 1
Images

滨水意向 2
Images

公园意向
Images

高架意向
Images

广场意向
Images

4.1.3

CITY OF MIRROR
镜像城市

学生：Yuan LU, Jiangning Sun, Xiaoyong ZHANG, Louis

小陆家嘴因为忽视城市文化传统，忽视人性而饱受争议，然而在城市的另一边，仅一江之隔的浦西地区，我们却很容易发现宜人的空间尺度，舒适的街道生活，传统的老城巷以及那些最代表上海的风俗文化。我们期待着两岸的对话，期待宜人的城市空间，期待着舒适的城市生活。

镜子在日常的生活中无处不在，从城市空间上的相互呼应到空间肌理的拓扑契合，镜像已然成为空间对话的一种有效的方式。在浦东与浦西的对话中，设计考虑从镜像的角度出发，试图发掘黄浦江两岸的必然与或然的联系，从而为提升整个小陆家嘴地区的城市生活品质找到适宜的方式。

浦西与浦东的镜像主要从物质和精神两个层面的城市更新上得以实现，物质层面，力求镜像浦西宜人的空间尺度，舒适的街道生活，连续的步行系统以及可达性较强的开放空间，从而构筑浦东空间品质提升的物质载体；精神层面，力求镜像浦西多种类、多层次的商业和娱乐氛围，持续的多种的城市活力，从而构筑浦东城市生活的软实力。只有综合了两个部分的镜像，才能使得城市与镜像的主题演绎的更加完美和交融。

镜像是与城市生活相关的一种现象，转换成城市设计的手段，我们试图从编织、融合、重构三个角度去解释经过镜像之后的城市空间形态究竟是怎样的。

Lujiazui is highly controversial because of ignoring the urban cultural traditions and human nature. However, on the other side of the city, Puxi area across Huangpu River has been very charming: the pleasant human-scale space rich street life, traditional Old Town as well as the most representative customs and culture of Shanghai. We look forward to cross-strait dialogue, pleasant urban space, and comfortable urban life.

MIRROR is everywhere in daily life. From the mutual echo of urban space to the fitting of the spatial texture, it has become an effective way of space dialogue. In order to establish the dialogue between Pudong and Puxi, the designers try to dig out the inevitable or probable links between both sides of Huangpu River to find the appropriate way to enhance the quality of urban life throughout the Lujiazui area.

Puxi and Pudong's mirror is mainly achieved from the urban renewal of the material and spiritual levels. In the material level, it strives for the pleasant spatial scale of mirror Puxi, comfortable street life, continuous walking system, and accessible open space, thus building material carrier to enhance the quality of Pudong space; in the spiritual level, it strives for diversity mirror like Puxi, multi-level commerce and entertainment atmosphere, and urban vitality, thereby constructing the soft power of Pudong city life. Only by combining the two mirrors, the theme of city and mirror can be interpreted in details.

Mirror is a phenomenon associated with the urban life. By converting it into a means of urban design, we try to explain the mirrored urban spatial form from the perspectives of weaving, fusion and reconstruction.

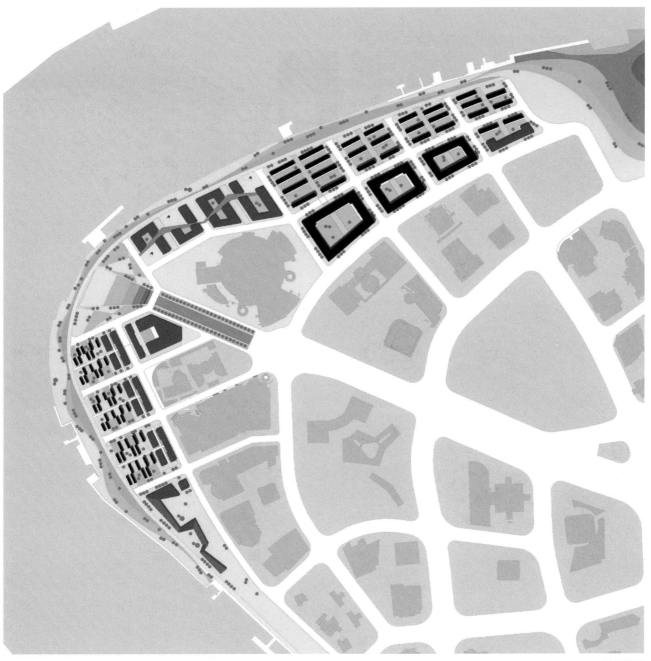

总平面图
Master plan

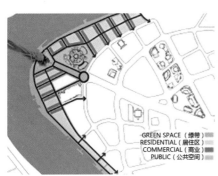

功能分析
Function

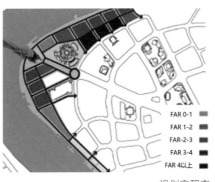

规划容积率
Planned FAR

车辆流线
Vehicular Movement

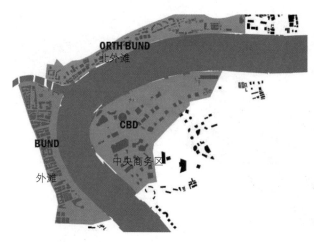
滨水区
Water Front Districts

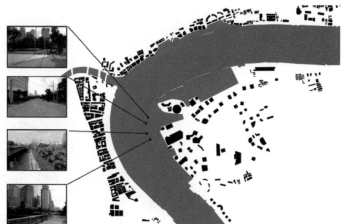
空白空间
Void Space

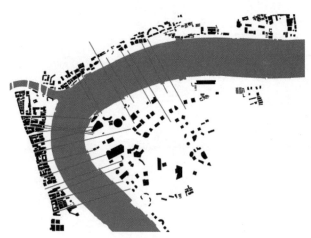
镜像反射框架
Reflected Framework

恢复期
Convalescence

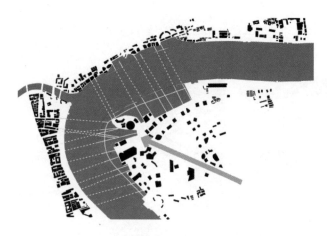
街区系统
Block Framework

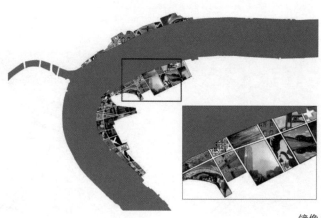
镜像
Reflection

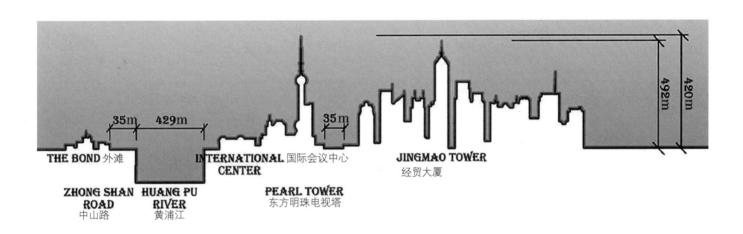

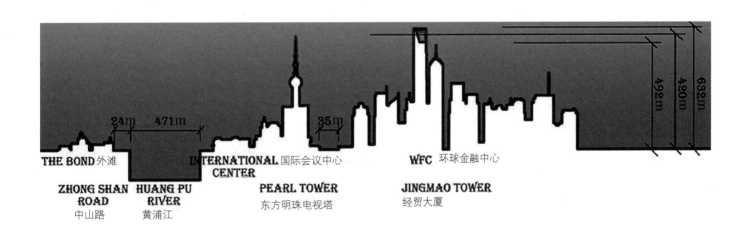

陆家嘴年代变化分析
Analysis

WATERFRONT VISION FOR CBD
镜像后的陆家嘴滨河区

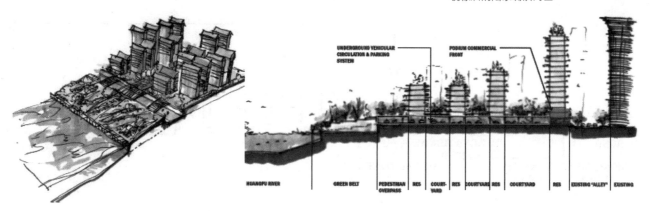

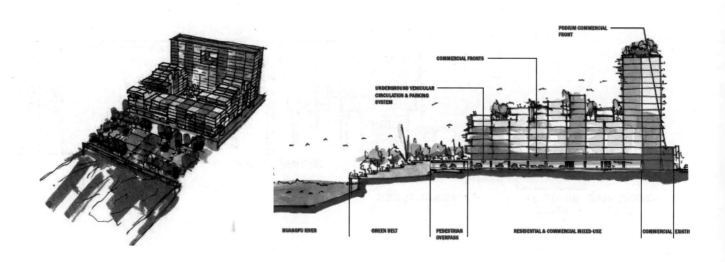

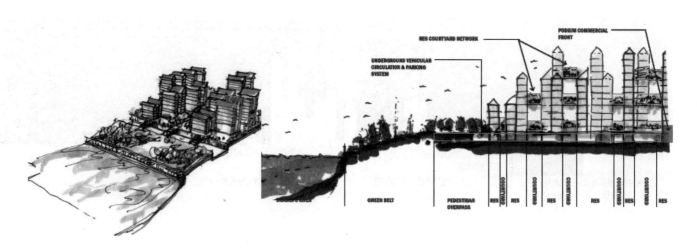

轴侧
Shaft Side

滨水建筑分析
Analysis

方案效果
Prospective

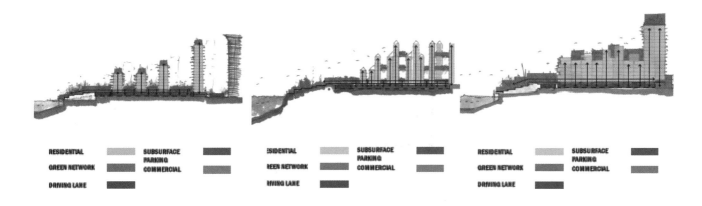

滨水地块功能分析
Function

4.1.4

THE PUBLIC DOMAIN & THE CITY OF MOVEMENT
公共空间与流动城市

学生：Biqing GE, Qin XIE, Huichao ZHU, Deanna Murphy, Paul Jones, Reggie Tabor

鉴于基地缺乏足够的公共空间、基础设施和充分的都市性，设计者提出了三条设计策略：一是对丢失的公共领域重新界定；二是提供多样的系统，提升环境品质较低的区域；三是提供新的基础设施，作为连接个体过于主导的私人领域和非常有限的公共领域的一个连续界面。具体而言，通过建立连续的路径，既能联系各个节点，又能作为一个公共和私密空间之间的界面，连接不同层级之间的交通流。

As Lujiazui lacks an adequate public domain, civic infrastructure and sufficient urbanity, the designers adopt three approaches. The first approach is to propose a way of occupying and defining a public realm lost in residual space. The second approach is to provide different systems that serve to urbanize a district without urbanism. The third approach is to provide new infrastructure that will act as a continuous boundary serving as an interface between an overly dominant private realm and severely limited public realm. That is, a continuous path that both connects nodes and serves as an interface between public and private will serve as sectional connections between layered movement systems.

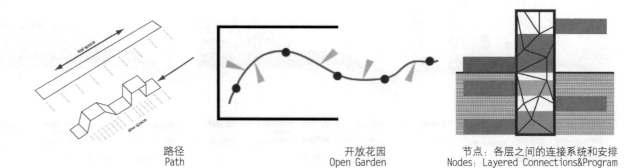

路径 Path　　　开放花园 Open Garden　　　节点：各层之间的连接系统和安排 Nodes: Layered Connections & Program

总平面图
Master plan

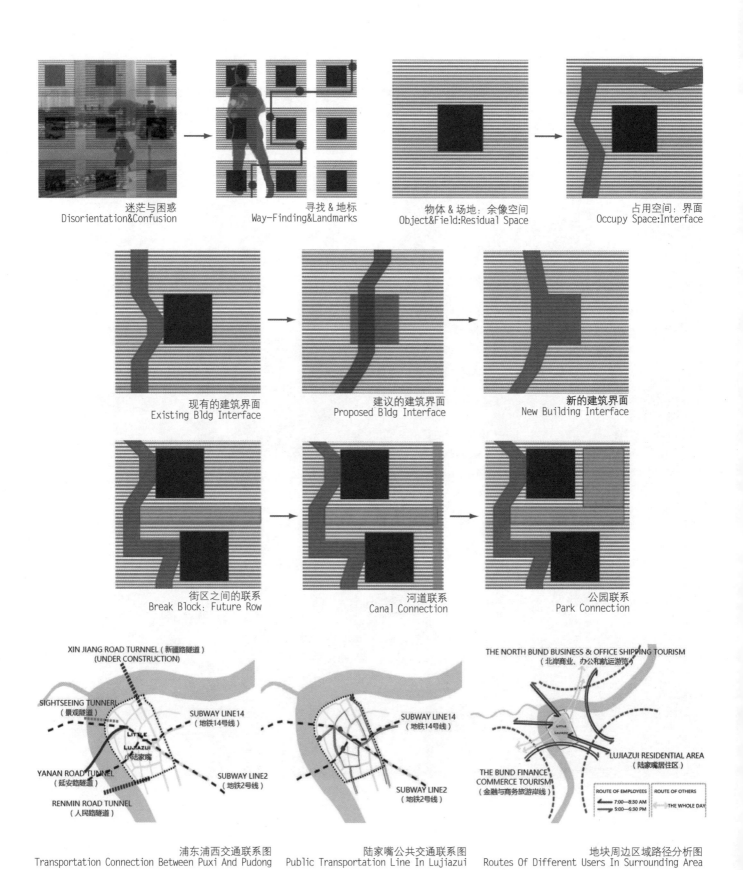

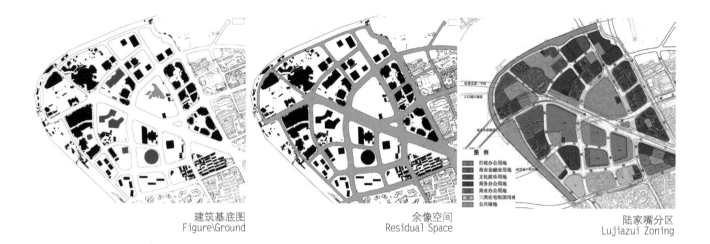

建筑基底图
Figure\Ground

余像空间
Residual Space

陆家嘴分区
Lujiazui Zoning

街道等级
Street Hierarchy

地铁线路（2 + 14）和车站
Metro Lines (2+14)&Station

公交线路
Bus Lines

人行道等级
Pedstrain Hierarchy

地下人行道
Underground Pedstrain Spaces

高架人行道
Elevated Pedestrain

69

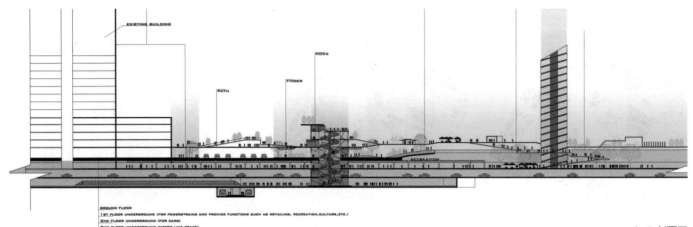

1-1 剖面图
1-1 Section

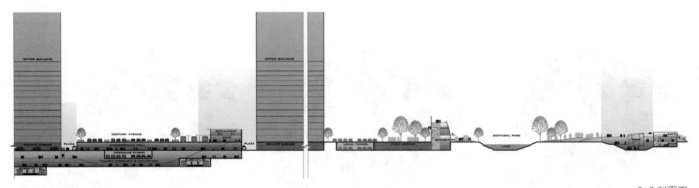

2-2 剖面图
2-2 Section

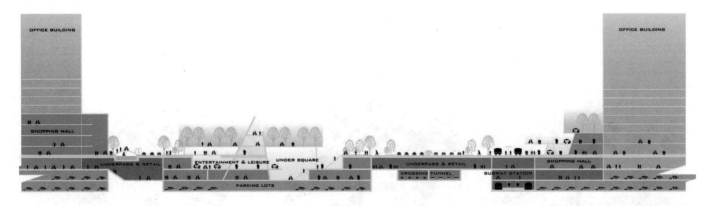

3-3 剖面图
3-3 Section

SWOT 分析
SWOT & PEEST ANALYSIS

	Stength 优势	Weakness 劣势	Opportunity 机遇	Threat 挑战
Political 政策	· Support from government · 政府扶持	· Administrative regionalization · 行政区划	· Two international centers · 两大国际化中心	· Other project competition · 其他项目竞争
Ecological 生态	· Lang waterfront and existing green · 长滨水岸线和现存绿地	· Negative microclimate · 消极微气候条件	· Govement's emphasis on ecology · 政府对生态的重视	· Vast construction · 大规模建设
Economic 经济	· Successful Financial centre · 成功的商业中心	· High run cost Benefit maximization not realized · 高运营成本会制约利润最大化	· Continuous investment on infrastucture · 持续的基础设施投资	· Investment on infrastructure is lower than other aspects · 基础设施的投资力度较弱
Social 社会	· Good working sites for elites · 精英汇聚	· Citizens have fewer choice to enjoy the facilities · 城市居民难以享受资源	· Emphasis on Face to Face communication · 强调面对面交流	· Market economy makes it ineversible (the gapbetween poor and rich) · 市场经济下不可逆的贫富差距
Technological 技术	· Modem materials · Sufficient Technical support · 现代材料； · 充分的技术支持	· Low efficiency · High cost for Maintenance · 低效率； · 较高的维持费用	· Positive ideas on application · More mature construction technology · 应用创新思维； · 更成熟的建设技术	· Promotion of modern technology is limited · 对现代技术的推进作用遭到限制 · It's difficult to combine the new kind of technology with the existing constructions · 难以将新技术与现有建设联系起来

路径意向 1
Images

路径意向 2
Images

4.1.5

PLAYFUL CITY PLAYFUL LIFE
乐活城市，乐活生活

学生：Howard Wang, Stuart Dryden, Minqing ZHANG, Yan CHAO, Yue AN

陆家嘴是一个充满了非凡压力的地方。在这里工作的商务人士很容易会陷入日常工作生活的巨大压力中。这里工作的人们已经远离了他们的童年和其他的美好回忆。忙碌的生活方式使得他们几乎忘记了游戏和猎奇的概念。他们似乎更像是二维世界里的小蚂蚁。

这个项目试图将严肃与活泼、高效率和完整性同时融入陆家嘴商业区。城市设计不应该试图控制使用者，而是应该为多样的、不可预知的事物提供更多的机遇。这个"游乐场"将为用户提供一个环境，在这里，使用者能够在独特的时间和地点体验独特的记忆。

Lujiazui is a place of extraordinary stress. It is a place where the business person can be easily caught up in the pressures of every day work life. The businessman is far removed from his childhood, a distant memory caught in another word. With his busy lifestyle, he has all but forgotten the concept of play and curious exploration. He is merely an ant in his two-dimensional world. This project attempts to juxtapose the serious with the playful while retaining the efficiency and integrity of the lujiazui business district. Urban design should not attempt to control the user but rather provide opportunities where multiple, unpredictable uses can occur. This "playground" will provide an environment where the user can experience unique memories specific to a unique time and place.

陆家嘴金融中心区在一天不同时段的人群活动特征
"Activity Trees" of Lu Jiazui Residents

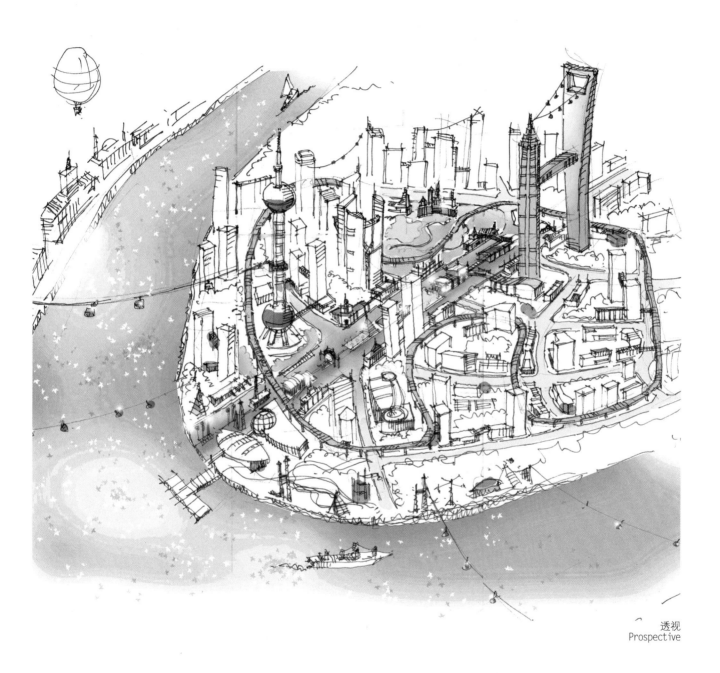

透视
Prospective

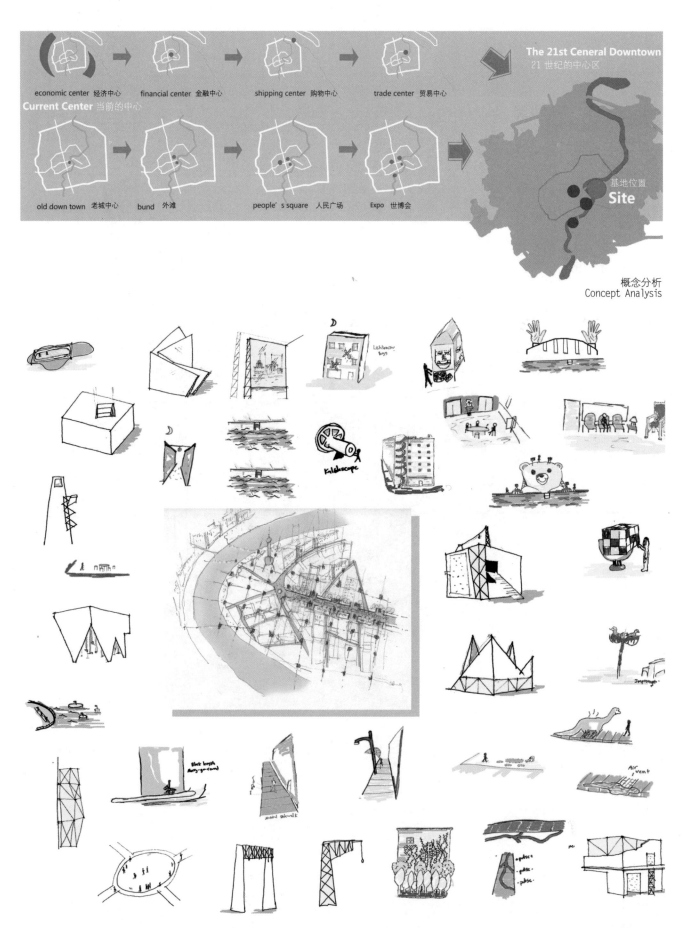

概念分析
Concept Analysis

方案概念分析
Concept Analysis

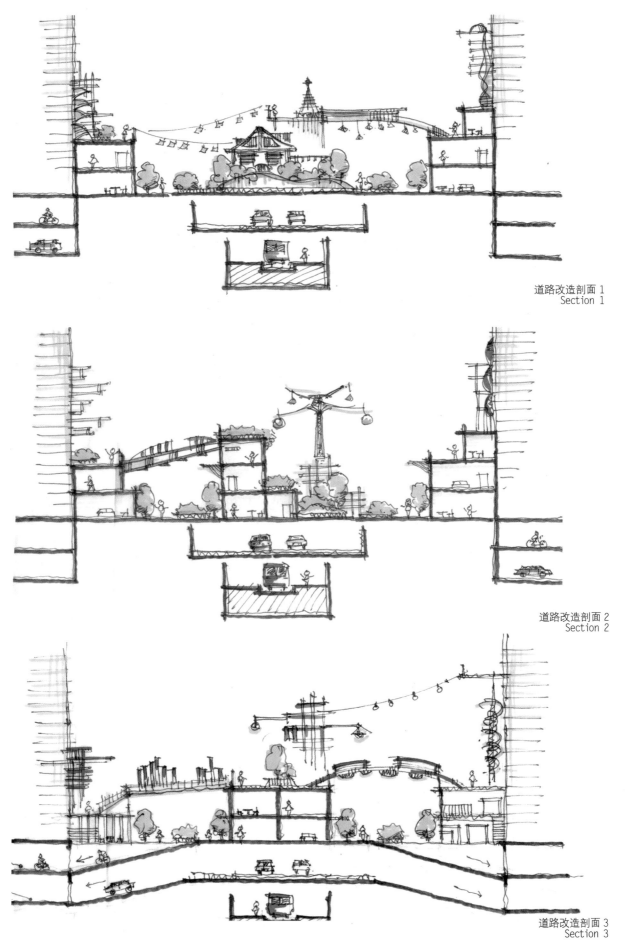

道路改造剖面 1
Section 1

道路改造剖面 2
Section 2

道路改造剖面 3
Section 3

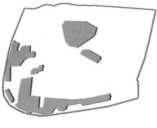
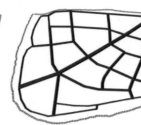
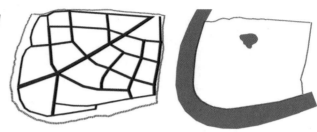

地标	绿化空间	道路网	水资源
Landmarks	Green Spaces	Road Network	Water

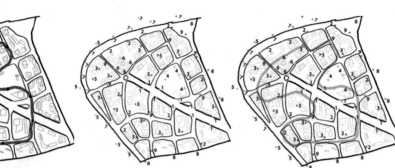
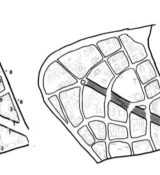

场地分析
Site analysis

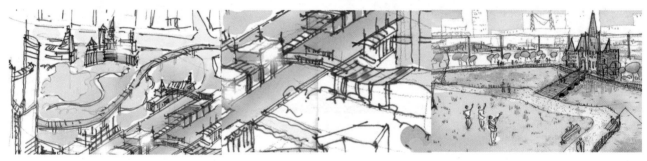

游乐场意向
Images

建筑意向图
Images

步道意向
Images

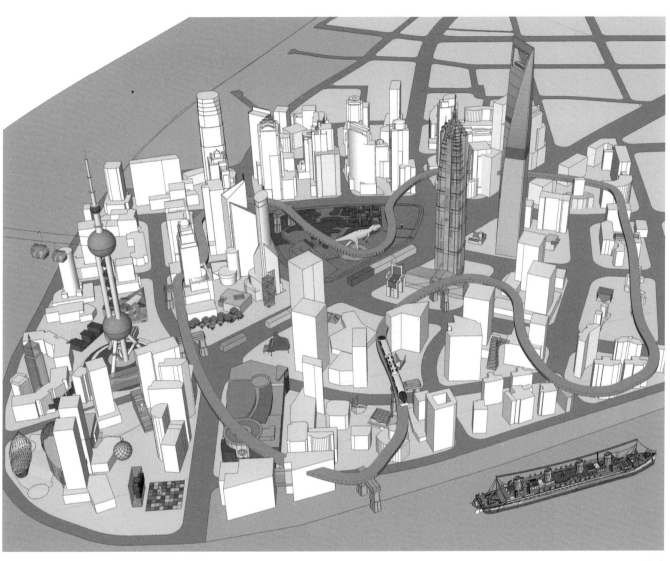

鸟瞰
Aerial View

4.2 上海金山区城市生活岸线概念性城市设计
CONCEPTUAL URBAN DESIGN OF COASTAL AREAS IN JINSHAN DISTRICT, SHANGHAI

指导老师：田莉，李晴
INSTRUCTORS: Li Tian , Qing Li

4.2.1 空间的变异——浮游的岛屿
SPACE METAMORPHOSIS: FLOATING ISLAND

4.2.2 有机都市主义
ORGANIC URBANISM

4.2.3 回归海洋
BACK TO THE SEA

4.2.4 彩虹带
LIVING FROM THE WALL TO RAINBOW

4.2.5 立体游戏
LEVEL OF GAMES

上海金山区城市生活岸线基地区位
Introduction to Jinshan Site

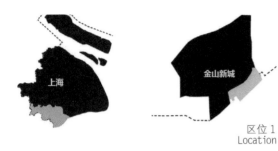

区位 1
Location

基地位于上海市金山区滨海新城，是金山区目前唯一的一段生活岸线，是上海最后的渔村——金山嘴渔村所在地，同时与上海最具海洋生态价值的所在地——方案目标：金山三岛海洋生态自然保护区隔海而望。

方案目标是将金山滨海区域打造为体现上海国际大都市滨海特色的重要"名片"。

The site is located in the coastal area in the Jinshan new town, Shanghai. This site is the only public shoreline in Jinshan District, and the last fishing village, JinShanZui fishing village, in shanghai is located in the southern part of the site. Jinshan Islands, natural ecological island faces the site across the sea.

The goal of this design is to become a landmark which is characterized by the coastal city image of Shanghai.

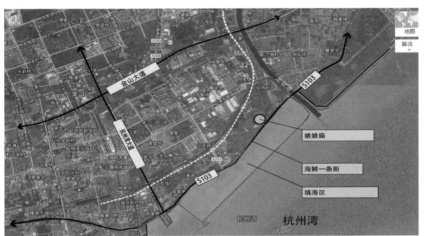

区位 2
Location

金山三岛之大岛
Pictures of the Site

基地范围
Site Boundary

渔村水渠
Pictures of the Site

南侧高层区
Pictures of the Site

地铁站
Pictures of the Site

4.2.1

SPACE METAMORPHOSIS: FLOATING ISLAND
空间的变异——浮游的岛屿

学生：Sravan Singh, Beatrice Lange, Viola, Josh, Teddy

水是这个基地的重要元素，也诠释了方案设计的自然本性。因此，我们从水的"三态变化"过程中找到设计灵感，水可以从冻结的冰到半固体状态再到液态，最后变成气态。这一过程可以用来隐喻从城市肌理过渡到大海肌理的这一过程，我们可以通过设计，来实现从城市到大海的平滑过渡。

填海造地的成本很高，对海洋生态环境的影响很大，而且还要面临洪水、海啸和台风的威胁。借鉴马尔代夫与荷兰填海的经验和技术，我们在方案中尝试设计了一种漂浮结构，不仅能够适应潮水的涨落，保持安全，而且漂浮的结构对海洋的生态影响也较小，关键是相较于填海造地来说，漂浮结构的成本非常低，可行性强。

The site can be visualized as a strip of land from the grid nature of the city to complete fluidity of the sea. The important element of the site is water that defines the nature of our project. Thus we derive from the various states of water i.e. from frozen ice to semi-solid to total liquid and then to gas. This is used as a metaphor of the relationship between city and the sea.

Land reclamation comes at a price. It has a huge ecological impact. The biodiversity of the sea is altered while the costs are too high for dredging and landfills. As displayed in the case of Maldives and Netherland, our plan making floating structures that adapt to the tides and are safe from flooding. Also since they are floating they have minimum impact on the sea. Costs for floating structures are comparable with land reclamations.

概念图
Concept

方案以金山火车站为中心，根据TOD的开发与规划策略，按照与火车站的距离远近将地块分成了四个区域，由内向外，首先是第一个区域，以高密度的商业开发和办公空间为主；接着是以混合功能、零售、博物馆、美术馆为主的第二区域；第三区域则是金山嘴老渔村，在保留渔村的基础上，主要进行居住的开发建设；第四个区域则是以木板路为主的动态变化的滨海开发区域，承担着文化走廊的角色。一条从城市开始的环绕基地一圈的连续的绿带，将基地内的空间开放空间、绿化空间都串联起来。

通过空间变异的概念，将城市和海洋有机地结合起来。从城市的网格状肌理到渔村的自由流线形态的道路，最后是充满流动性的漂浮岛屿。

方案模型
Prospective

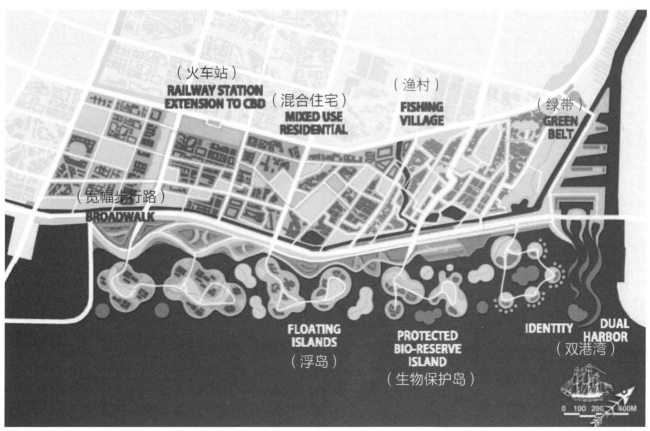

总平面图
Master plan

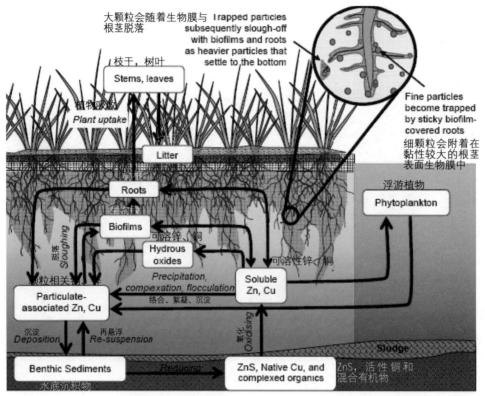

浮岛生态概念图
Eco-conception of Floating Island

本案例研究了一些列位于海边受到洪水威胁但生态环境却保持得很好的案例。马尔代夫，一个印度洋岛国，以优美的海洋度假环境而闻名全球。但随着全球气候的变暖，海平面上升使得马尔代夫在不断地失去陆地。因此，他们提出了漂浮岛屿的办法来应对。其他类似的案例还有著名的阿联酋城市——迪拜；荷兰的阿姆斯特丹，这个城市全年中有将近1/3的时间面临着洪水的威胁；中国海南的漂浮城市；还有韩国首尔的全世界最大的人工漂浮岛等。

In Netherland's planning and design of floating land has becomes mainstream since global warming might bring too much water for dykes to keep off.

Some municipalities like Amsterdam and Rotterdam include floating land in their spatial development plans.

Most projects are still small scale and depend on land based infrastructure.

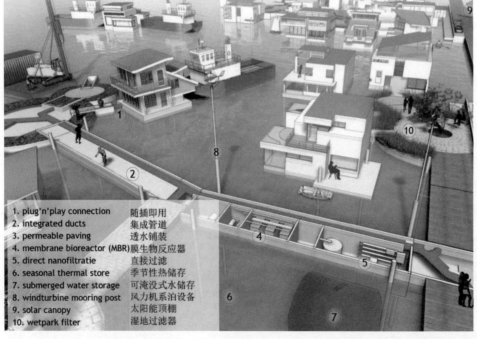

1. plug'n'play connection　随插即用
2. integrated ducts　集成管道
3. permeable paving　透水铺装
4. membrane bioreactor (MBR)　膜生物反应器
5. direct nanofiltratie　直接过滤
6. seasonal thermal store　季节性热储存
7. submerged water storage　可淹没式水储存
8. windturbine mooring post　风力机系泊设备
9. solar canopy　太阳能顶棚
10. wetpark filter　湿地过滤器

漂浮建筑物及其支撑性基础设施
Floating Building and Their Infrastructures

场地参观和调查表明，金山嘴地区拥有一个非常重要却脆弱急需得到保护的生态环境。因此，我们要把规划和设计对生态系统的影响程度降到最低，并建立和金山新城之间的关系。

新的多样化的活动项目给本地区带来新的旅游机会。整个场地根据一个大的概念为主统一规划，并以一种友好的可持续发展的方式与城市连接。

Site visit and research made obvious that the environment of Jinshan Zui is vulnerable as well as important and rich and needs protection. The approach is to plan and design with a minimum of intervention in the ecosystems, and the area needs to be improved and connected back to the Jinshan city.

New and diversified development of activities leading to Tourism opportunities in the area. The entire site is planned as per a larger concept and is then reintegrated to the City through Metamorphosis Sustainable initiatives for an Eco-friendly city.

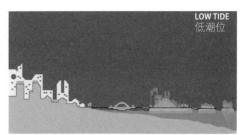

水位变化
Alteration Of The Sea-Level

沪杭公路意向
Images

滨水广场意向
Images

公共空间意向
Images

鸟瞰
Aerial View

Investigating the fishing village, ecology and landscape
渔村、生态与景观调研

Converting the fishing village into a built up plan
渔村建成环境转换

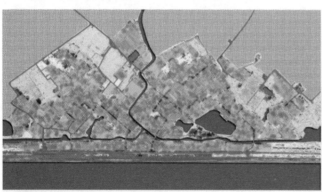

Dividing the village into neighbourhoods (urban blocks) based on our Metamorphosis concept
基于蜕变的概念，将渔村划分为多个社区（城市街区）

Inclusion of green space to balance built and un-built & also inward looking blocks
基于绿地空间平衡建筑与非建筑空间，引入景观斑块

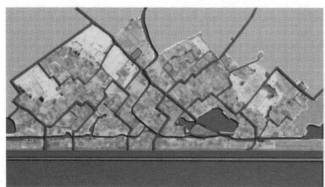

Services and Connections bring the blocks together from suburban spontaneous settlements to urban neighbourhoods
将郊区自发居民点转换为城市社区

New development now undertaken in the urban framework of the village, thus retaining the essence and yet regenerating
在城市街区框架下，村庄得到了新的发展，同时保持了自身的特质并得到再生

渔村更新改造
A Revitalization Of Fillshing Village

道路系统分析
Road Hierarchy

公共空间－私人空间
Public-Private Spaces

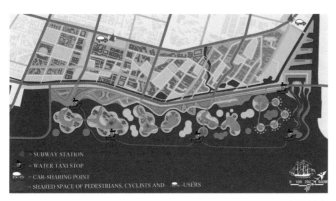
交通站点
Traffic Station

滨水景观系统
Waterfront Landscape System

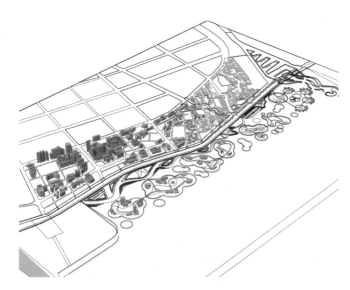

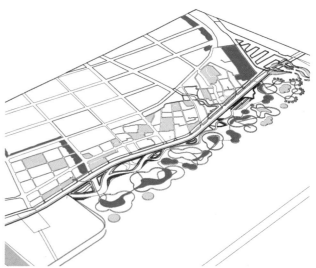

4.2.2

ORGANIC URBANISM
有机都市主义

学生：Emilie Schmitz, Jessica Steee-Hardin, Armin Mehdipour, Anna Symonenko, Francesca Achilli

有机都市主义是指通过充分考虑场地特征，使基地上的建筑和周围环境成为一个相互统一、相互联系的整体，进而促进人类居住环境与自然之间的和谐共存。

首先，有机都市主义的概念性想法是绿色有机的，需要将水、能源等要素整合起来。基于这个策略我们将会在设计中提供较多的绿化空间和滨水空间，并注重能源的可持续利用；然后，我们开始关注基地的形状。我们发现，它是一种柔和自由的波浪形，具有类似海浪的特征。地块的形态好像要张开怀抱，去拥抱海上的三个岛屿。因此，我们在设计中尝试拉近三个岛屿和基地之间的距离，使得金山三岛不再那么"尽在眼前，却遥不可及"。最后，我们在设计中强调要保持渔村的完整性，包括保护建筑和周边的自然生态环境。同时，增加适当的旅游和商业，来带动当地的经济发展，提高村民的收入和生活质量。

The concept of organic urbanism is to promote harmony between human habitation and the natural world through a design that approaches the site sympathetically and is well integrated with its site that buildings and surroundings become part of a unified, interrelated composition. We started the design process with identifying conceptual ideas, site axis and shape, and finally site program and design. First, the conceptual ideas are green, water, energy, integration, and organic. Then we moved to the shape of the site, which is a soft boundary that is influenced by the perpendicular axis the site shares with to the nearby three islands. The shape of the site appears as if it is reaching out to hug the islands. Finally, the program and design kept the fishing village intact.

概念图
Concept

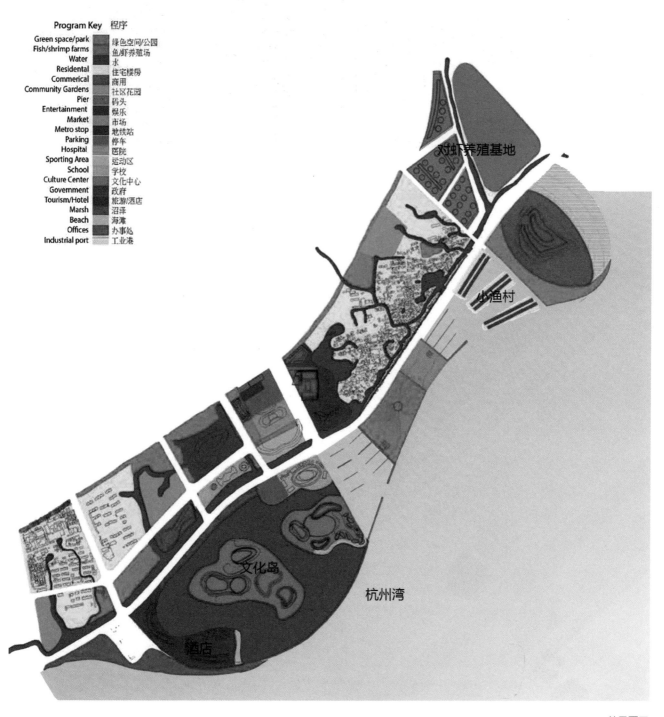

总平面图
Master plan

现状照片
Present Situation Picture

有机村庄的概念就是通过将建筑和周边的环境完美融合在一起的相关设计方法，创造出人类聚居地和自然之间的和谐状态。

The concept of Organic Village is to promote human habitation and the natural world through a design that approaches the site sympathetically. This is well integrated with its site that buildings and surroundings become part of a unified interrelated composition.

我们观察到通向三个岛屿的轴线。
We observed The Site's Axis To The Three islands.

然后，我们将海浪模式和轴线结合起来。
Then We Integrate The WavePattern Research And The Axis To The Three Islands.

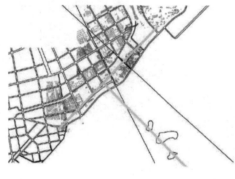
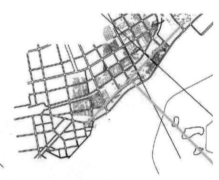

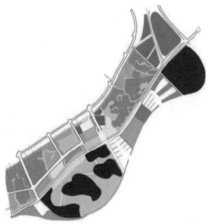

土地开发示意图
Sketch Map Of Land Development

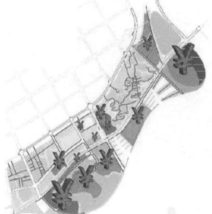

土地价值示意图
Land Value Sketch Map

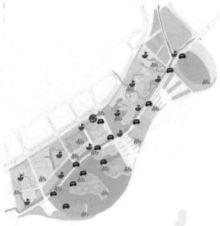

交通示意图
Traffic Sketch Map

ROAD DIAGRAM
道路示意图

道路分析
Road Map

剖面关系分析
Section Cuts And Diagrams

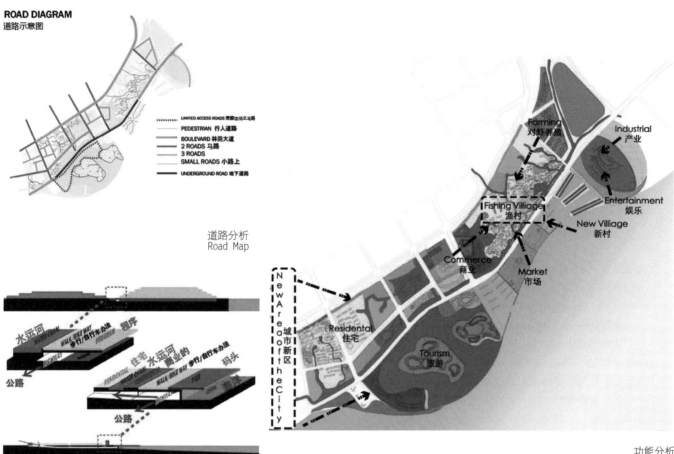

功能分析
Function

退后三维模型
3D Model

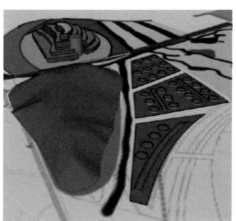

对虾养殖基地空间意向
Images of farming

文化岛意向
Images of cultural island

4.2.3

BACK TO THE SEA
回归海洋

学生：Tatiana Mukhina, Tina, Chuk, Luca, Long YIN

海，是什么？
它如何影响我们的生活？
它叙述着怎么样的故事？
又带给我们多少不同的感受？

在我们看来，我们的基地相较于欧洲、美国以及中国香港等地的海边渔村来说，完全没有体现出临近大海的优势——临海却不"近"海，金山嘴渔村与大海被一条防波堤生生割裂开来；没有因为临海而带来优美的生活环境；海岸线没有得到很好地利用，现状处于荒废状态；老渔村百废待兴……到底是什么原因导致了这一系列事与愿违的尴尬？

为了重新点燃海边渔村的活力与生机，我们从宏观、中观和微观三个尺度上对这一问题进行了深入的研究。首先，通过了解金山地区的历史变迁和海岸线变迁，我们抓住了潮水涨落这个突破点。其次，两次现场实地调研让我们对渔村和海洋的隔离有了非常深刻的认识，得出了必须打通渔村连接海洋的视线通廊的重要结论；第三，我们针对地块内的居住者和游客进行了问卷调查，使用问卷调查中反映出的数据来指导我们的设计。

最终，我们提出了我们的设计理念——BACK TO THE SEA 回归海洋，包括三大设计原则：潮水、时间和视线，通过这三个要素的营造来达到回归海洋的目的。

The sea... What is it? How does it influence people lives? What narratives and stories does it tell? What feelings does it create?

Comparing to similar places 'fishing-village-at-the sea' in the world (Europe, US, Hong Kong, etc), we find that the Jinshan site has not taken advantage of its natural and cultural resources. Therefore, based on a deep research of issues of different scale: macroanalysis, mesoanalysis and microanalysis, we conducted interviews on the site with residents and visitors. Interviews shown that JinShanZui has much potential to be developed, and on the other hand there are problems to be solved. Finally, we establish three main principles: Tide. Time. View.

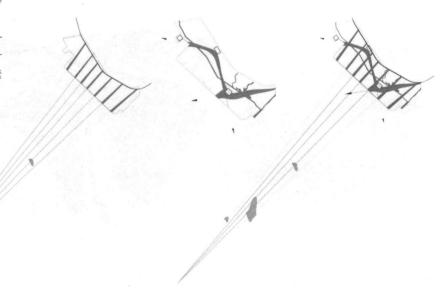

方案概念
Concept

View:

The approach is cutting channels within the entire area. There are two types of channels: the first type consists of three broad channels with starting points on the railway station – the starting point for visitors. Two channels are cut in the directions of two nods – superstructure and floating market. Third channel starts from the point of intersection of the metro line and site boundary. It leads to the floating market. These three channels create a main public space in the whole area. The second type of channels is a straight channel facing the sea, oriented to the islands, and starting from the points, where city road system intersects with the boundary of the site, and we replace roads with channels.

视线廊道：

我们在整个基地范围内营造出两种类型的水道，来打通基地和大海、各类重要节点之间的视线通廊。第一类从西往东依次连接海边构筑物、火车站、漂浮市场、龙泉港北部入口等重要节点，以火车站和漂浮市场为折点，形成"之"字形的东西向水道，并将地块划分成五大组成部分；第二类则是面向大海的直线水道，延长线汇聚于金山三岛的一点。

The coastline along Jinshanzui is not used efficiently. Natural residences can make it more attractive & offer a view to the islands.

金山嘴海岸线没有得到有效的利用。自然住宅能够使海岸线看起来更加具有吸引力，还能提供看海、岛屿的视线。

The view from the village to the sea is unfavorably blocked by several buildings. Only small building structures along the coastline are tolerable to secure the view.

几幢建筑挡住挡住了村子通往海的视线。因此，沿着海岸线，只能建设小体量的建筑，防止遮挡视线。

Unused roads can be transformed into water canals and provide a distinctive alternative to public transportation.

不适用的道路可以转化成水道，提供更具特色的公共交通。

设计策略
Design Strategy

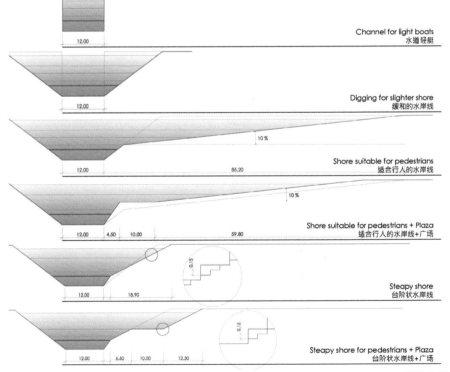

潮汐分析
Tide Analysis

Tide:

We want to bring water inside the area with the canals. Dogged land can be used for the reclamation area. Canals are to be connected with the sea and bring sea water inside. Embankments along canals are designed to emphasize this level change. Steps and walking paths are designed on different levels to be flooded during high tide and to work as public spaces during low tide.

潮汐：

我们设想通过人工的运河将海水引进地块内，这样就能够在地块和大海之间建立联系。沿着这些运河，设计不同剖面的河堤形式，来强调因潮水涨落而带来的景观和功能上的不同效果。比如在不同高度上设置台阶和步行道，在高潮位的时候被淹没，而在低潮位的时候则可以成为步行的公共空间。

Time:

To give a life to the site of Jinshan for a long period of time, we established a system of nodes. Buildings, which stand on important intersections such as channels and boulevards. The lifetime of these buildings is much higher than the life expectancy of average buildings. With appropriate material, these buildings should last for several centuries, and they can be used as cultural, emotional or institutional facilities, for example, hospitals, theaters, governmental buildings, tourist attractions, schools and landmark buildings. The nodes system also includes a superstructure located near the city beach.

This superstructure should function as a landmark building lasting for more than 1000 years as Gyze Pyramids in Egypt or Great Wall of China. We want it to remind about this site many years forward, even when the whole site will be filled with water. The superstructure is a construction that works as a lighthouse, but functions also as a vertical cemetery of buildings. When the buildings on the site are demolished, they fill the superstructure with supercompressed materials, and it grows.

时间：

为了创造出金山嘴地区在时间上的延续感，我们建立了一个节点系统。我们在人工运河水道、林荫大道等的十字路口处放置代表文化、情感的核心建筑，如医院、剧院、政府大楼、学校以及地标性建筑等。这些建筑拥有更长的寿命，能够延续上百年，甚至更长的时间。在海岸边，我们特意设计了一座巨型构筑物。这个构筑物不具有实际的生活功能，但它可以作为一个灯塔，也可以作为建筑物的"垂直墓地"，意思是当基地上的其他建筑物拆除时，可用它们的材料来建造这个构筑物。这样一来，这个构筑物能够随着地块内建设的变迁而逐步升高，时刻提醒我们这里的过去，即使是这个地块被海水淹没了，这个构筑物仍然能露出水面，延续上千年的时间，像埃及的金字塔和中国的长城一样，留下时间的记忆。

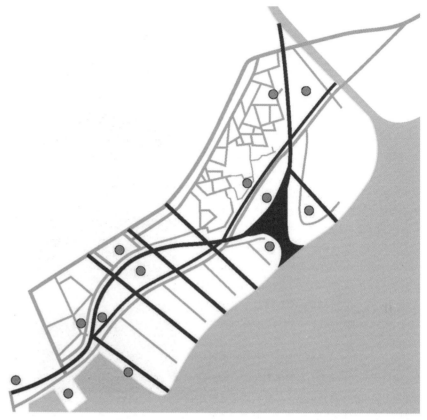

节点系统
Node System

Light shines into the room,
Waking him up
He smiles
A strong cup of Longjing
And the sounds of the street
Remind him of his youth.

He returned home yesterday.
75 years is no joke.
The hospital has been here a long time.
Almost as long as he.
Most buildings around here
Didn't last as long.
They were built up
During a time of success
And showing off.
Continuously re-built
And crafted a strong place in time.
The superstructure by the bay is almost finished.
A monument of the past
And grand ideas.

Where are they now?
Those who remained,
The kids always left
To see the world,
On the move
To catch the next big thing.
They'll learn someday.

He goes to meet his grandson
On the old-fashioned train from Shanghai.
They embrace, and look out towards the water.
"Grandfather, I am so glad to be back"

阳光照进房间
轻轻地，将他唤醒
睁开双眼，嘴角轻轻上扬
一杯提神的龙井
一阵嘈杂的汽笛声
唤起年轻的记忆

他昨天才回
在75岁的年纪
迎接他的，
是这里一座几乎和他一样年纪的医院
其他的建筑却没这么好运
在被一次次地建造着
炫耀着
持续性的重建
精雕细琢出这样一个地方
海岸边的建筑即将完工
一座承载过去和未来的纪念碑

他们在哪里？
那些曾经的人儿？
孩子们以这里为起点
去闯，去向外面的世界
在追寻的路上
一直前行……

但终有一天
他会欣喜地去接上他的孙儿
在那节老式的、从上海开来的车厢里
老少相拥
望着那一顷碧波
终于回来了，我的大海

Back to the Sea

The Questionnaire Survey 市民意向调查：

What are advantages and disadvantages if JinShanZui Village?
你觉得金山嘴的优势和劣势是什么？

What in your opinion is missing in JinShanZui Village?
你觉得金山嘴还缺点什么？

Can you name 3 aspects which you wish for the future of JinShanZui?
你希望金山嘴以后变成什么样子？

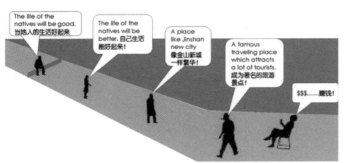

What do you think is the main function of the coast of JinShanZui?
你觉得这边的海滩该怎么利用？

After interview, we have focused on four specific target groups. The interviews conducted in the fishing village serve as a basis. The landscape and infrastructure are mainly designed according to the activities, values and interests of these target groups. Target groups are: fishermen, Shanghai playboys, tourists & families, businessmen.

问卷调查之后，我们聚焦四类目标群体，分别为渔民、上海有钱人、游客以及商业人士。地块的功能分区、景观设计、基础设施配置都是根据四类人群的活动、价值取向和兴趣来安排的。

We divide the site into 5 areas: high-density mixed-use area, old fishing village, marina and resort, eco-park, and green and public space island.

我们将地块分成五个区域：高密度的混合功能区，金山嘴渔村，海边度假区，生态公园区，绿色公共岛。

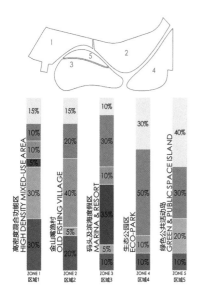

ZONE 2
Old Fishing Village
区域2: 金山嘴渔村

FISHERMAN
渔民

MOST IMPORTANT ACTIVITIES
最重要的活动

- Fishing 捕鱼
- Running gastronomy 饮食业
- Public facilities 公共设施

FAVORITE LOCATIONS
最喜欢的区位

- Harbour 港
- Floating Market 漂浮市场
- Fishing Village 渔村
- Administration 行政

OLD FISHING VILLAGE
15%
20%
40%
5%
20%

ZONE 3
Marina and Resort
区域3: 码头及滨海度假区

SHANGHAI PLAYBOY
上海花花公子 (上海有钱人)

MOST IMPORTANT ACTIVITIES
最重要的活动

- Leisure, entertainment 休闲娱乐
- Shopping 购物
- Yachting 帆船运动

FAVORITE LOCATIONS
最喜欢的区位

- Luxury hotel, panoramic view 有良好滨海景观的奢侈酒店
- Marina 码头
- High-density urban area 高密度城市地区

MARINA & RESORT
10%
30%
10%
35%
5%
10%

ZONE 4
Eco-Park
区域4: 生态公园区

TOURIST FAMILY
游客

MOST IMPORTANT ACTIVITIES
最重要的活动

- Leisure, nature 休闲、旷野
- Shopping 购物
- Dining 饮食

FAVORITE LOCATIONS
最喜欢的区位

- City Beach 城市沙滩
- Eco-Park 生态公园区
- High-density urban area 高密度城市地区
- Fishing Village 渔村

ECO-PARK
30%
50%
10%
10%

ZONE 5
Green and Public Space Island
区域5: 绿色公共活动岛

FISHERMAN
渔民

MOST IMPORTANT ACTIVITIES
最重要的活动

- Fishing 捕鱼
- Running gastronomy 饮食业
- Public facilities 公共设施

FAVORITE LOCATIONS
最喜欢的区位

- Harbour 港
- Floating Market 漂浮市场
- Fishing Village 渔村
- Administration 行政

GREEN & PUBLIC SPACE ISLAND
40%
30%
20%
10%

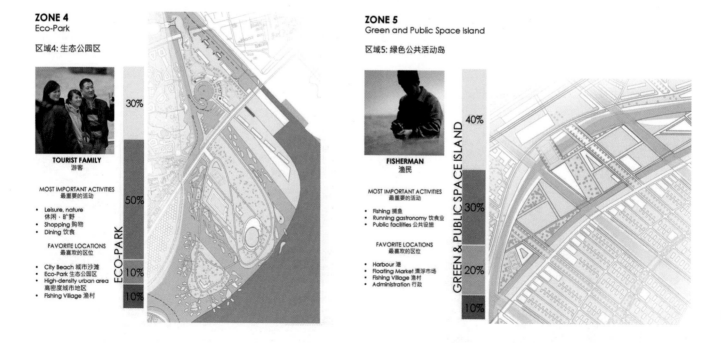

总平面图
Master Plan

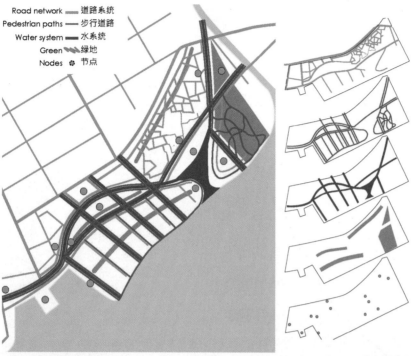
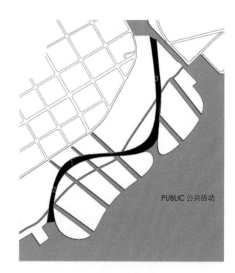

系统－节点
Layers-Nodes

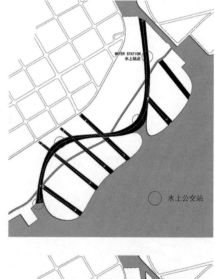
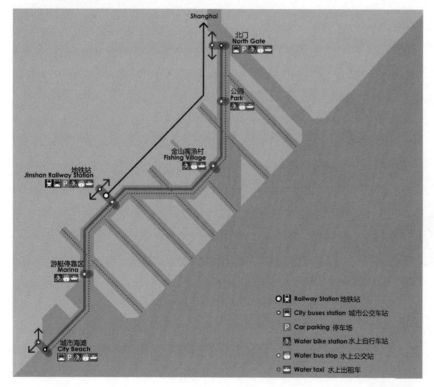

交通系统
Transportation System

水道功能
Function of Canal

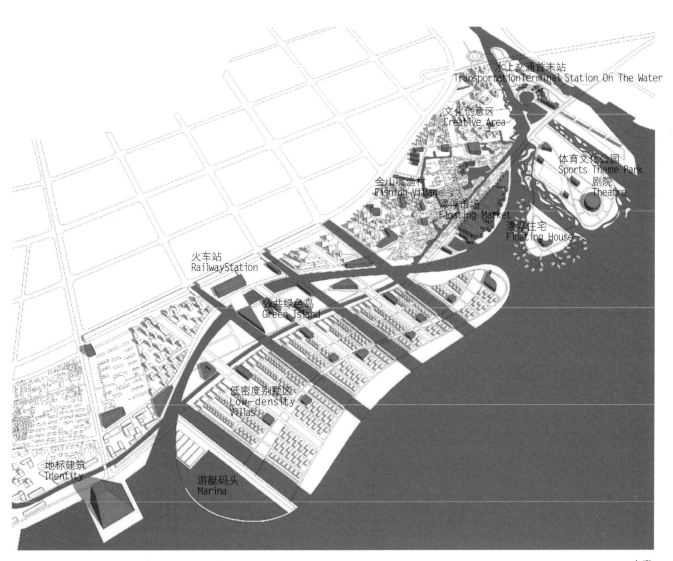

鸟瞰
Aerial View

4.2.4

LIVING FROM THE WALL TO RAINBOW
彩虹带

学生：Benedetta, Eveyn, Kiran Jayakumar, Oscar Manuel Cid Esc, Stefanie Fritze

方案的设计概念是"彩虹带"。彩虹带是我们这个设计的主题：首先，彩带联系了许多公共活动功能，比如剧院、水族馆、海鲜美食节广场、捕鱼活动区和金山历史露天展览区等，在东侧港口处收尾。

其次，彩虹带的带面是灵活变化的，通过高度变化使它适应现状建筑，新建道路可以从彩带的下方穿过而不至于切断彩带。沿着彩带边缘，我们还设置了有轨电车和人行步道，使彩带真正成为人们娱乐、休憩的公共场所。

The ribbon is the design concept of our project: it contains numerous public functions such as an open theater, an aquarium, an area for the sea food court, an open exhibition that guides the people to the museum of fishing activities and history of Jinshan, and it ends in the port. The surface of the ribbon has a dynamic identity: the height changes adapting itself to the preexisting elements. The new roads do not cut the ribbon surface but they pass under it, and the existing one are covered by the ribbon. Along the edges there is a tram line and pedestrian paths.

1. The wall
墙面

2. The wall as a barrier: only one access to the sea
墙体阻隔，只有一条亲水路径

3. Breaking of the wall: more access to the sea
打破墙体，创造更多路径

4. Orientation towards the island
导向三个岛屿

5. Identification of the main attraction point
确定主要吸引点

6. Connection with the ribbon
由彩虹带串联

概念图
Concept

最初听到金山嘴渔村，我们都以为渔村和大海之间会有非常紧密的联系，头脑中浮现出一幅海边渔村欣欣向荣的唯美画面。但是，当我们到达项目所在地时，现实情况却出乎所料，我们根本无法靠近大海。因为视线被防波堤挡住了，这堵"墙"从东到西，横穿过我们的地块，并且把老渔村和填海区生生划分开了。

针对现状问题，我们提出了四种策略：

Jinshanzui fishing village is supposed to be closely connected with the sea. However, when we arrived at the site, we found that the seaside was inaccessible in the site due to the block of the jetty. This wall crosses our site from East to West and divides the old fishing village from the new reclaim area.

In the design, we set up four strategies:

STRATEGY 1: redefine the concept of the wall as a link between the old and new urban texture.
Starting from the idea of continuity of the wall, we transformed it as a new identity of the place changing its shape and function making it the junction point between the three different parts.

策略1：重新定义墙（防波堤）的概念。墙可以是城市新旧肌理之间的链接，防波堤本身是连续的，我们保留它的连续性的特点，但改变它的形状和功能，使它成为连接金山嘴渔村和大海的纽带。

STRATEGY 2: creation of new axis and a new-strong infrastructure system.
We break the wall in different specific points in order to create the missing link using roads and pedestrian paths.

策略2：创建新的轴线和完善的基础设施系统。
我们在不同的特定点打断墙，引入道路和步行系统，从而横向连接渔村和大海。

STRATEGY 3: renovation of the existing houses and requalification of the river banks.
We think it is very important to renovate the whole village, not only the main road, paying attention especially to the dirtiness and pollution.

策略3：更新现有住房和提升河岸品质。
我们认为这是非常重要的是更新，不仅仅是主路，而是整个村子，特别要重视村庄脏乱的治理。

STRATEGY 4: ribbon as the concept for a new urban design.
The best shape to unify all these strategies is the ribbon: it becomes shape, expression and identity of the new urban texture. It is a colorful place with important public functions.

策略4："彩虹飘带"作为新的城市设计概念。
彩虹飘带状的形状能够很好地实现以上策略：它将很好地诠释新的城市肌理的形状、特征。这将会是一个多姿多彩的地方，同时又具有重要的公共职能。

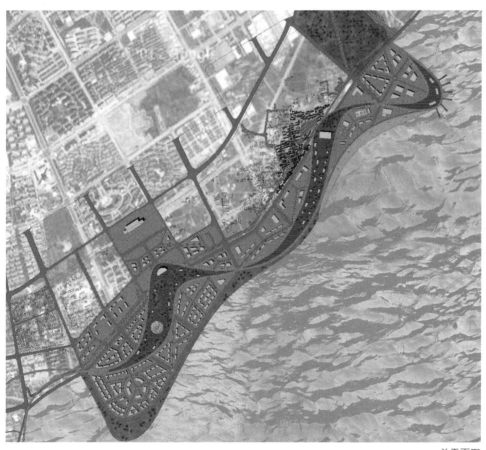

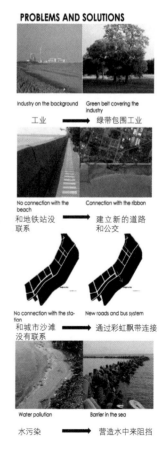

总平面图
Master plan

According to the needs and the natural resources of the area we design the functional program. This is not a restricted division but only the main zoning.

In the southern part of the site, easily connected to Jinshan city we placed the residential and commercial area as the main sources of money.

In the middle part where are fishing facilities and market, didactic activities for children and some hotels for tourists since the proximity to the ancient village.

In the northen part we design a sports area, a new beach for the entertainment of Shanghai inhabitants, a natural landscape connected to the natural park and the new port. In order to preserve the identity of the environment, the high rise buildings are situated in the inner part and the low rise along the coast.

The new green system invades the whole site in order to create a new vitality and attract more people. In conclusion, our design goal is not create a high density district without a specific identity, but a new one for people, their activities, their culture and their history.

根据地块内各地区的现状和需求，我们进行了混合功能分区。南部易于与金山市区联系，设置了居住区和作为主要经济来源的商业区；中部靠近金山嘴渔村，在这里设置了一些垂钓设施和渔市以及为孩子准备的教育活动场所，为游客服务的酒店等；北部则是港口区域，包括一个新的海滩、自然公园和龙泉港畔的重要港口。

总的来说，这个方案不是为了建造一个毫无特色的高密度街区，而是一个为了人、生活、文化和历史而建的新社区。

功能布局
Function

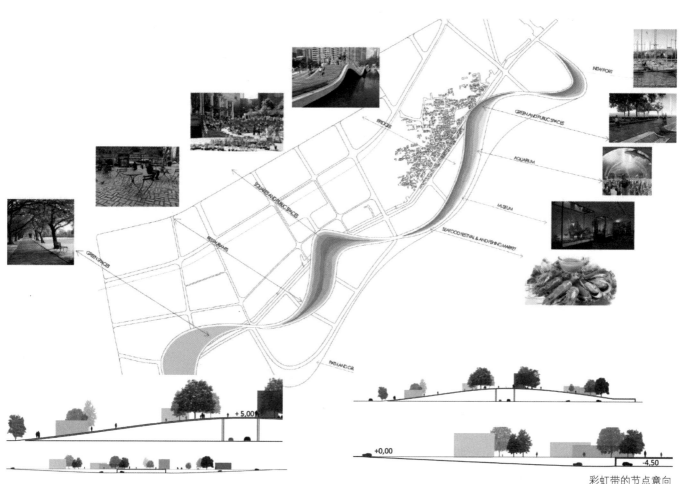

彩虹带的节点意向
Images Along the Ribbon

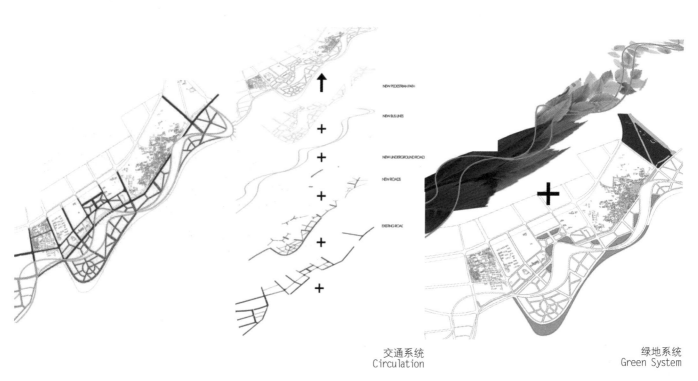

交通系统
Circulation

绿地系统
Green System

4.2.5

LEVEL OF GAMES
立体游戏

学生：Luca, Bruneaus, Grace, Luck

经过现场调查和初步的认知之后，我们对基地有了两个印象，第一印象是基地紧邻大海，却没有因为大海的存在而使得基地更加充满生趣；第二印象则是基地内存在着不同高度的变化，海平面、沪杭公路、金山嘴渔村等都处在不同的高度上。高差的变化带来了许多问题，如视线遮挡、人的可达性问题以及隔离等实际存在的问题。如果不解决因高差变化带来的负面影响，基地将很难发挥出原本近海应该有的优势。

因此，我们提出设计理念——"Level Game"，即立体游戏，希望通过我们的设计，创造出充满乐趣、丰富的高度变化，使基地焕发出生机。首先，在大海和渔村之间设计连接平台，跨过道路，直接连接，让人们能够便捷地到达大海；其次，通过加宽道路，植入多样的功能，变换高差来使道路充满步行乐趣和活力。

Goal: To make JinShanZui a modern coastal town in the southern part of Shanghai, a special cultural and tourist attraction center.

After the site survey, we have got two impression on the site. The first impression is the inaccessible seaside. The second impression is the variety of heights.

In order to achieve our design goal, first of all we create intermediary platforms, which can connect the water front and the village, and proviside access to other facilities in the site. Secondly, the level game, for instance, "Live" street is designed to increase accessibility as well as a providing mixted functions at different level.

概念图
Concept

问题聚焦：

环境污染；

植被、少量树木和动物的生态隔离；

渔业缺乏动力；

城镇文化以及遗迹破坏；

水体未得到保护；

垃圾无处理堆放。

PROBLEMS CITED

Pollution;

Extinction of vegetation, rare plants and animals;

Population pressure;

Lack of "Engine" in the fishing industry;

Destruction of cultural sites and towns.;

Unprotected water bodies;

Poor disposal of waste.

RECOMMENDATIONS

To make full use of the transportation system and green space;

To blend the built environment and the surroundings and promote investment;

To preserve the three islands and maintain the culture of fishing;

To preserve and conserve cultural and ancient sites;

To estabish new relations;

To give new values;

To improve the sense of community;

To find old traditions for a new economic development;

Development of agriculture and fishing actives;

Improvement of mobility system;

Creation of new connections with the new and old part;

Promoting tourism;

Creation of a new view;

Creation of attractive points.

建议策略：

充分利用交通系统；

充分利用绿地空间；

建筑环境与自然环境的融合；

促进投资；

保护三个岛屿；

延续"渔"文化；

保护地方文化和历史遗址；

创建新的关系；

赋予其新的价值；

增强不同区域间的联系；

寻找古老的传统；

找到一个新的经济开发点；

发展农业和渔业；

发展移动设备；

在新与旧之间创造新的连接点；

促进旅游业；

创造新的景观；

创造新的吸引点。

现状
Today

未来
Tomrrow

概念深化（Level Ideal）

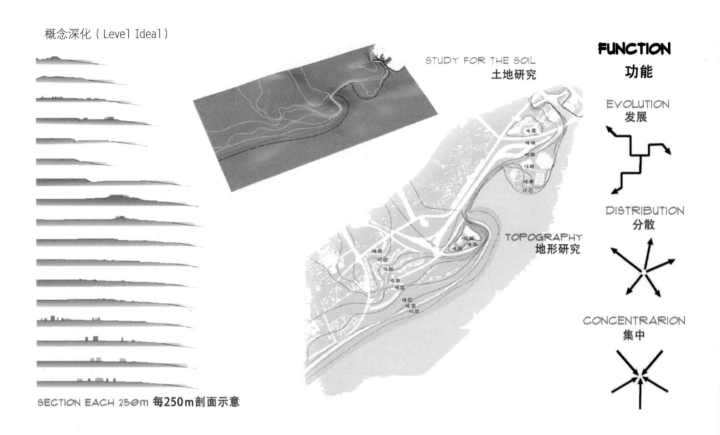

SECTION EACH 250m 每250m剖面示意

STUDY FOR THE SOIL
土地研究

TOPOGRAPHY
地形研究

FUNCTION
功能

EVOLUTION
发展

DISTRIBUTION
分散

CONCENTRARION
集中

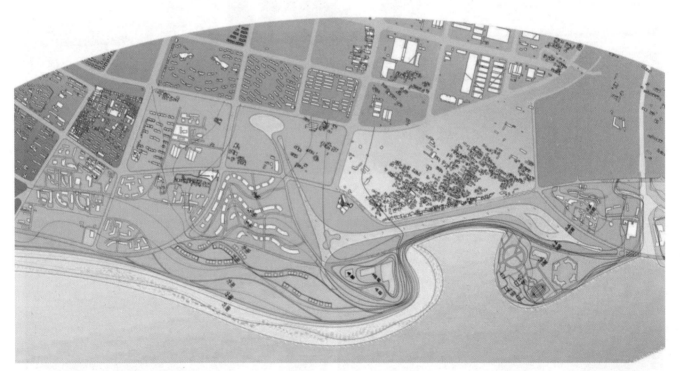

总平面图
Master plan

总结：

通过上述策略，我们希望创造一个融合"创新"与"传统"的旅游文化中心。

这将有助于凝聚金山地区成为一个整体，实现综合性的经济社会发展，改善居民的生活条件，协调"人"与"环境"之间的关系。

城市规划应该考虑过去、现在和未来的人们的需求，瞄准更有价值的目标，实现可持续发展。

CONCLUSION

With the above approach in our plan, we expect to have a special tourist and cultural center, blending the new and the old.

This will help the people of Jinshan to achieve a comprehensive economic, social development goal. This will improve the people's living condition and acquire the balance between human-being and environment.

Urban planning should put into consideration the past, present and future generation. This should aim at better valuable results for sustainable development.

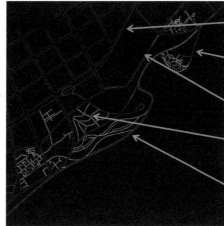

道路等级
Road Hierarchy

- Public Transportation 公共交通道路
- Main Trunk 主干道
- Underground Passage 地下通道
- Secondary Road 次级道路
- Pedestrian & Cling Path 慢行道路

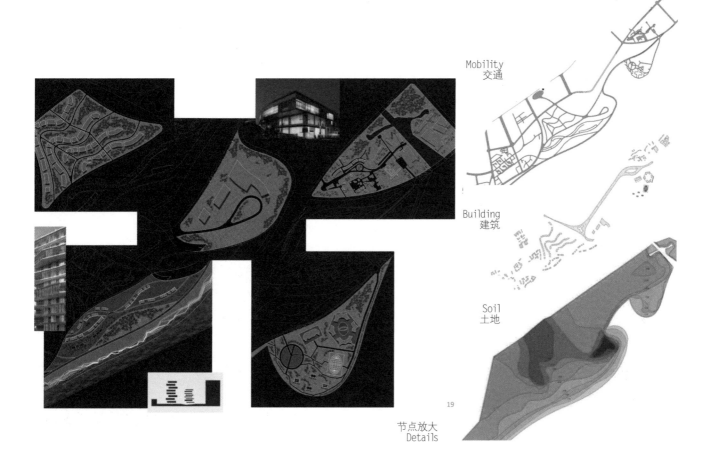

节点放大
Details

Mobility 交通

Building 建筑

Soil 土地

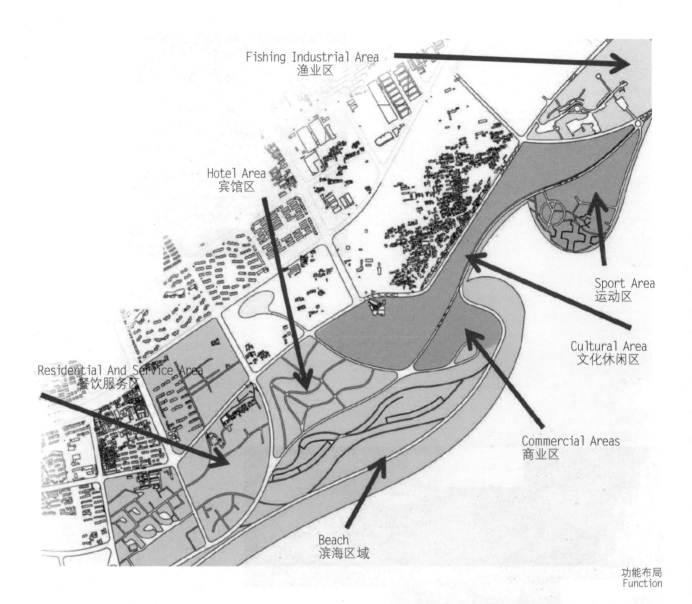

功能布局
Function

海滩意向
Images

体育区意向
Images of sports activity zone

滨海公共区意向
Images

滨水公共区意向
Images

4.3　上海金山区亭林镇中心区城市设计
URBAN DESIGN OF CENTRAL AREA OF TINGLIN TOWNSHIP, JINSHAN DISTRICT, SHANGHAI

指导老师：田莉，李晴，陈竞姝
INSTRUCTORS: Li Tian, Qing Li, Jingshu Chen

4.3.1　水岸生活
WATER SIDE LIFE

4.3.2　蜕变
METAMORPHOSIS

4.3.3　很亭林
THAT'S TINGLIN

4.3.4　植绿亭林
TINGREEN

亭林基地区位
Introduction to Tinglin Site

亭林镇位于上海市金山区东北部，是金山区距中心城区最近的城镇。亭林历史悠久，早在4000年前，即有先人在此活动，民国时期是浦南三大商业重镇之一。基地位于亭林老镇中心，面积约64公顷，其内水系纵横，并有读书堆、古松园等历史遗迹。

Tinglin town, located in the northeast of Jinshan district, is the nearest town to Shanghai central city in Jinshan. Tinglin has a long history. Ancestors had lived here as early as 4000 years ago. It used to be one of the three major commercial towns in Punan district in the Republic of China Period. Located in the center of Tinglin old town, the site, with an area of 64 ha, has a complex river system and many cultural relics such as Reading Hill and Pines Park.

现状用地
Land Use

建筑高度分析
Building Height

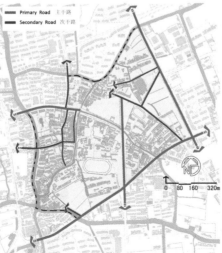

交通系统分析
Road Hierarchy

现状照片
Picture of the Site

WATER SIDE LIFE
水岸生活

学生：Song Chen, Zhen Wang, Edoardo Mezzadra, Sadeghi Naser, Agostini Gaia, Marina Parlapiano

本方案以"水岸生活"为设计主题，探索沿河生活模式，通过连续的滨水空间改造与功能更新，打造"天蓝水青、古镇古香、多元多彩"的亭林水生活长廊。

历史重现：以人文水生活为主线，打造多元化的历史文化活动空间。

探寻时尚：将洁净河流与滨水活动结合在一起，形成生态、宜居、富有活力的现代生活水岸。

闲暇时光：针对基地内部大量老年人和儿童，设计休闲生活空间，融合宜居、生态及多元化思想。

This proposal takes "water side life" as the design theme, explores the pattern of water side life through continuous water-side space transformation, and functional updates. We expect to create an image of "blue sky and green water", an ancient town with vintage finds and colorful lives of the water pavilion living gallery.

Historical Life

The design follows the intellectual life with water, and creates multinational and historical cultural activity.

Modern Life

It combines the clean rivers and activities around forms a modern waterfront with ecological, livable and energetic environment.

Leisure Life

The design concentrates on thinking about the abundant aged and children inside the site to design casual living, which brings together the ecological, livable and multinational thoughts.

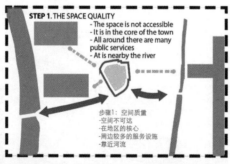

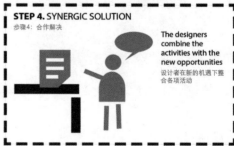

方案概念分析图
Concept Analysis

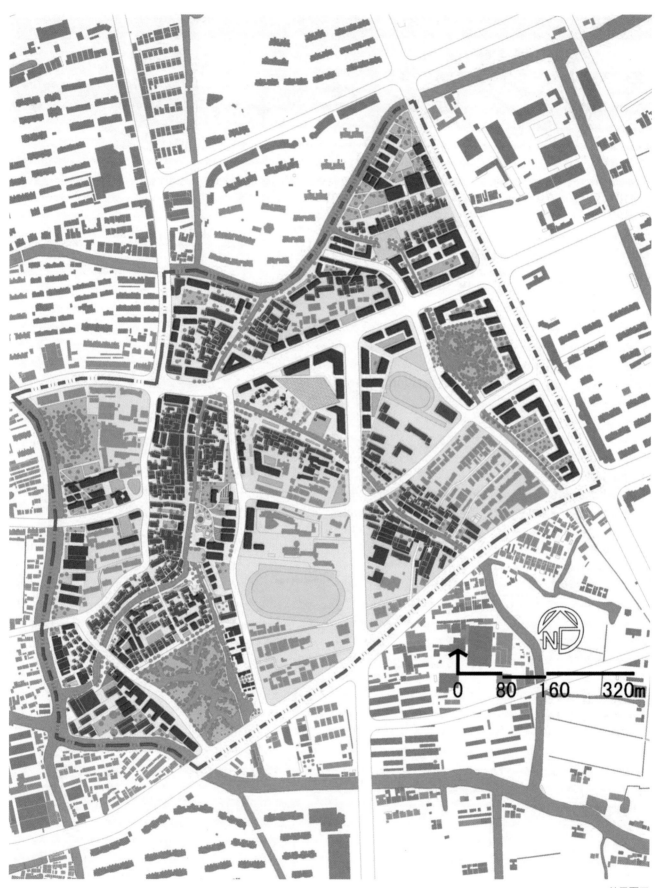

总平面图
Master Plan

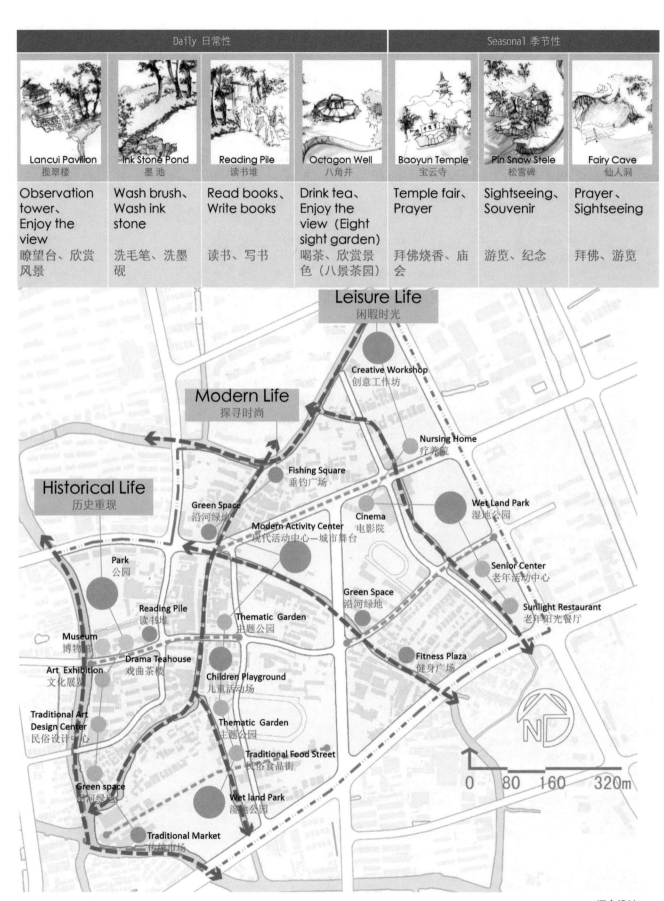

概念设计
Concept Design

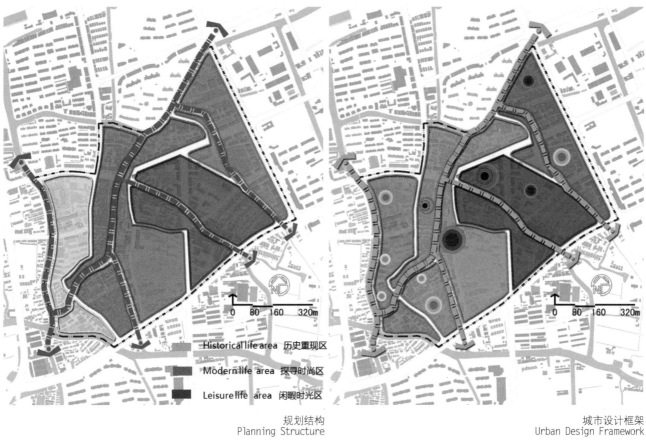

规划结构
Planning Structure

城市设计框架
Urban Design Framework

道路等级
Road Hierarchy

步行网络
Pedestrian Network

113

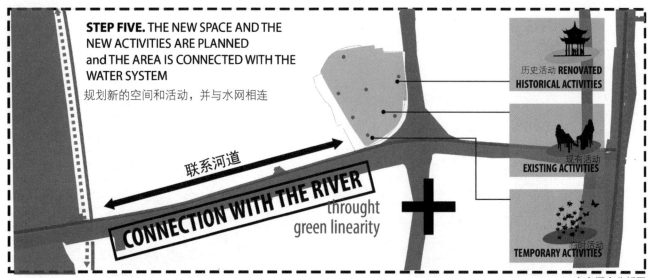

方案概念分析图
Concept Analysis

Historical Life : Traditional market
历史重现：传统集市

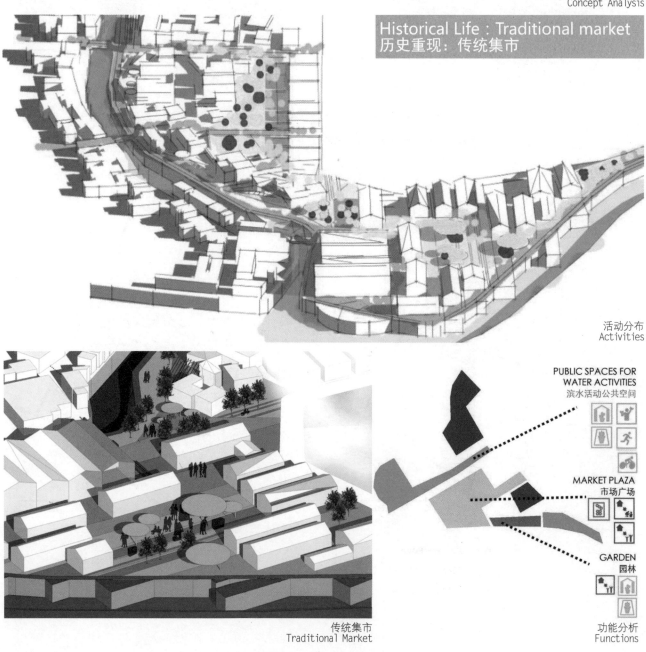

活动分布
Activities

传统集市
Traditional Market

功能分析
Functions

Modern Life : Temporary plaza
探寻时尚：季节性广场

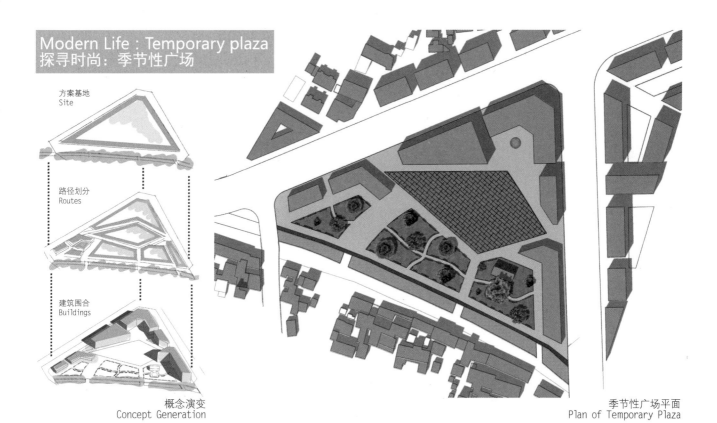

方案基地 Site
路径划分 Routes
建筑围合 Buildings

概念演变
Concept Generation

季节性广场平面
Plan of Temporary Plaza

Leisure Life : Cultural plaza
闲暇时光：文化广场

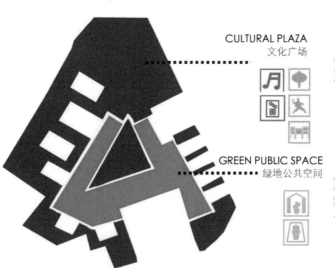

CULTURAL PLAZA 文化广场

GREEN PUBLIC SPACE 绿地公共空间

功能分析
Functions

绿色通廊
Green Path

文化广场意向
Cultural Plaza

中心广场
Central Plaza

4.3.2

METAMORPHOSIS
蜕变

学生：Rebecca Matilde Brollo, Eileen Chang, Matteo Fiorani, Maghsoudi Zahra, Xiaodong Pan

本项目以"蜕变"作为规划理念，希望在城市转变的过程中，变好的不仅仅是空间和土地，人同样可以在这一个过程中获得质变。通过对人群需求的分析，设计学习的空间、就业创业以及提升生活品质的空间，为人群提供提升自身、获取工作以及享受生活的机会。亭林镇具有水乡风格，基地内的河道贯通成网。基地内现有公园一处，主要的绿地空间为农田。自然元素、河流和绿地系统在亭林镇中扮演了重要的角色。这些元素共同创建了一个将各个地区衔接的网络系统。沿河将会植入很多重要的功能，例如创意水岸，生态市场和培训中心。绿地系统（由公园，城市园林和其他绿化空间构成）将会成为居民运动，农业体验和畅快游戏的重要空间场所。

In the process of urbanization, people as well as space are changing. Our topic is about how people are changing? How they change better? The transformation of Tinglin will be made possible by improving and accentuating the existing functions by providing more educational opportunities, new enrichment spaces and better employment prospects. The approach is demographically oriented and the plan began by examining the needs of the different people and designing the space for the community to learn, practice, and enjoy life in the city.

Tinglin has a strong character of water town, and a water network which has been formed for a long time. Besides, there is a park and green farmland in the site. The natural elements, the river and the green system play an important role in the town. Those elements together create a system that connects all the new areas. Along the river will be located many relevant functions such as the creative industry, the bio market and the training center. The green system is composed by parks, gardens and green areas where people can practice sports, farming and play freely.

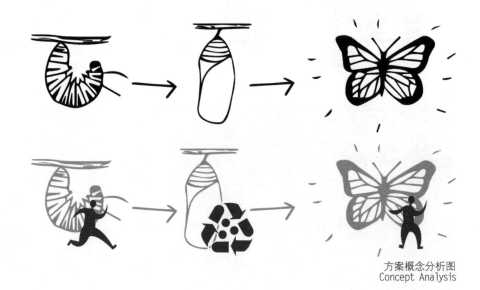

方案概念分析图
Concept Analysis

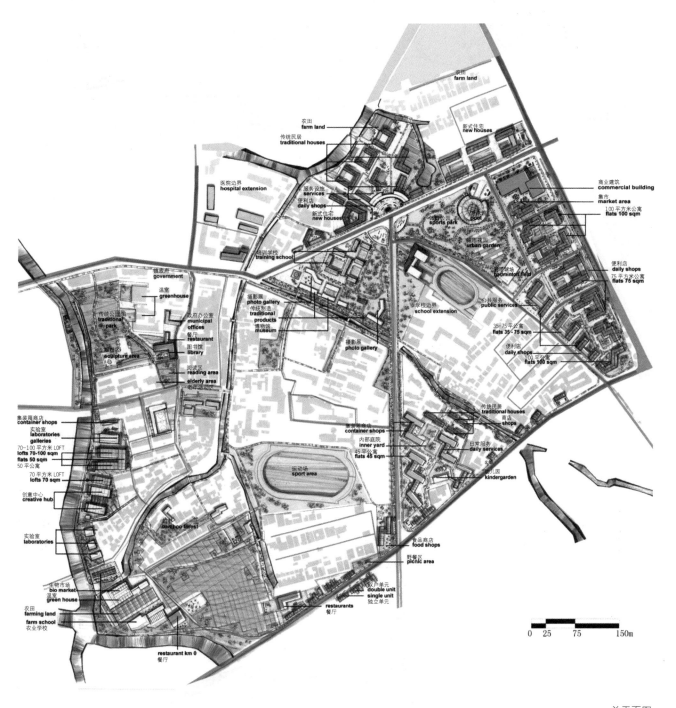

总平面图
Master Plan

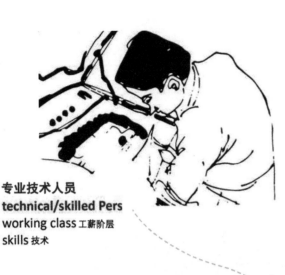

专业技术人员
technical/skilled Pers
working class 工薪阶层
skills 技术

无业/失业人员
unemployed-retired
poverty level- middle class 贫困阶层-中产阶层
farming, exchange products and house owner
农业、产品交易和房主

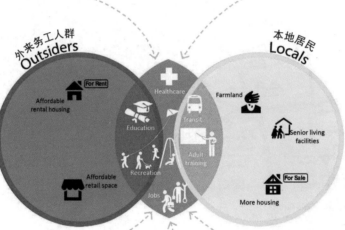

销售人员
sales persons
working class 工薪阶层
clothes and food
服装和食品行业

白领阶层
white collars
middle managers of enterprises 企业中层管理人员
construction materials industries 建材行业

个体经营者
owners of
private enterprises
中产阶层 middle class
餐饮与商店 restaurant and shops

现状人群及其需求分析
Current situation and need analysis

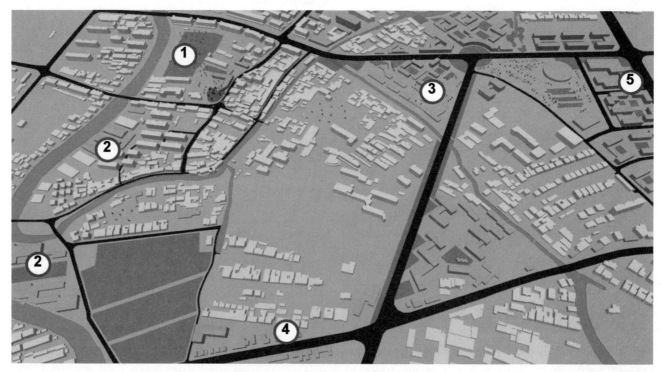

总体鸟瞰
Bird's-eye View

1 Cultural Center: Learning from the Past
文化中心：从历史中学习

2 Creative Banks: Learning from Practic
创意水岸：从实践中学习

图书馆意向
Library

水岸意向 　　　　　　　　　　水岸剖面
River Bank　　　　　　　　　Section of River Bank

文化中心意向
Cultural Center

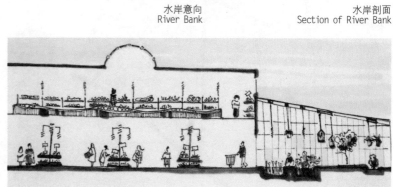

创意中心剖面
Section of Creative Hub

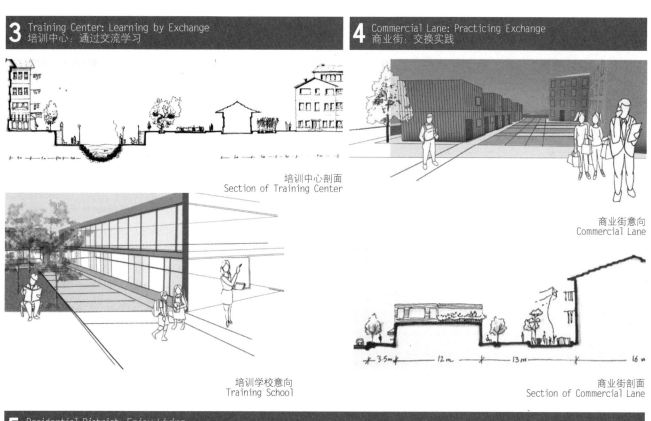

3 Training Center: Learning by Exchange
培训中心：通过交流学习

培训中心剖面
Section of Training Center

培训学校意向
Training School

4 Commercial Lane: Practicing Exchange
商业街：交换实践

商业街意向
Commercial Lane

商业街剖面
Section of Commercial Lane

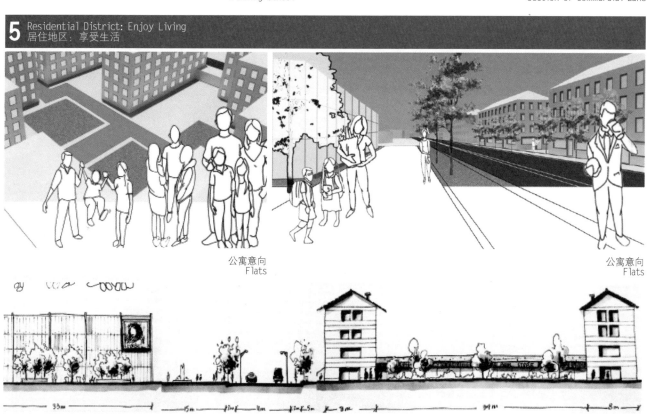

5 Residential District: Enjoy Living
居住地区：享受生活

公寓意向
Flats

公寓意向
Flats

商场与公寓剖面
Section of Shopping Mall and Flats

4.3.3

THAT'S TINGLIN
很亭林

学生：Bicheng Zhang, Biere Tristan, Kaban Duygu, Luca Manca, Mustafa Obaid, Sadeghi Zeynab

"很亭林"的主要概念是关注亭林老城的公共空间，其界定了亭林的主要特征，也是亭林的主要优势。"很亭林"是建立在亭林的历史文化特征上，如亭林的公共精神与持续存在的历史景观。"很亭林"的目的在于保持历史要素，赋予其新的功能与内涵。

亭林的各要素，如台、梁、水界面、街道等，在塑造亭林的空间品质上具有重要作用。通过提升这些空间要素，有助于建立更宜居的同时居民引以为豪的社区环境。我们的目的在于保持这些要素，使其重新整合至现代生活，通过振兴亭林的历史文化，并使其与居民的日常生活密切联系，避免历史文化环境的商业化或僵化衰退。

The main concept behind "Hen Tinglin" is to concentrate on the communal spaces that form the main strength of the area, reflecting its history and identity. "Hen Tinglin" aims not only to preserve the historic elements, bur rather present them in a new light, adding a new, optimistic layer.

Tinglin's typological features—its quays, bridges and waterways as well as its streets and open spaces—are playing the key role in Tinglin. Revitalizing those spaces by emphasizing all the elements that represent Tinglin will help building a better environment and a stronger communal identity that the people are proud of In this project our aim is to represent the historic elements, keeping its essence and reintegrating them within the modern city scape of Tinglin. Thus, we will help reviving Tinglin's historical background and connecting it with resident's everyday life without leading to the commodification and mummification of the area.

总平面图
Master Plan

亭林历史分析
History of Tinglin

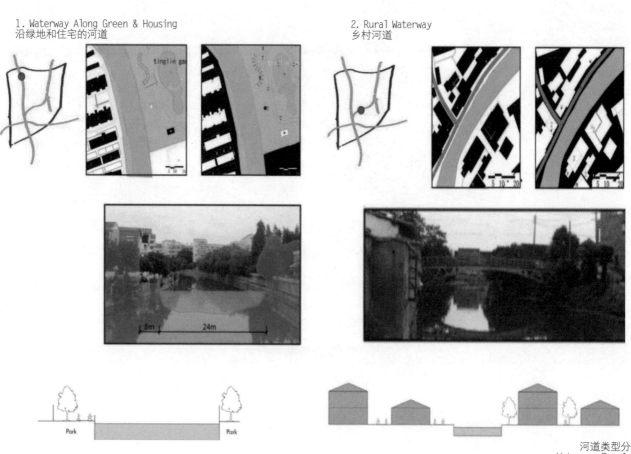

河道类型分析
Waterway Typology

1. Main Commercial Road
主要商业道路

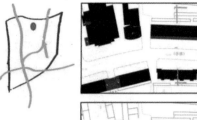

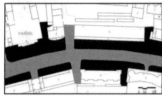

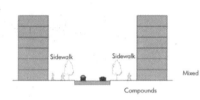

2. Regional Road
过境道路

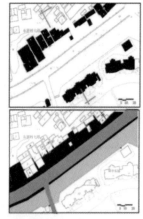

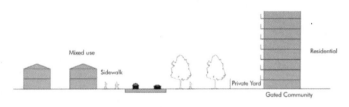

3. Rural Street
乡村街道

4. Traditional Waterfront Street
传统滨水街道

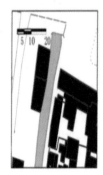

道路类型分析
Roads Typology

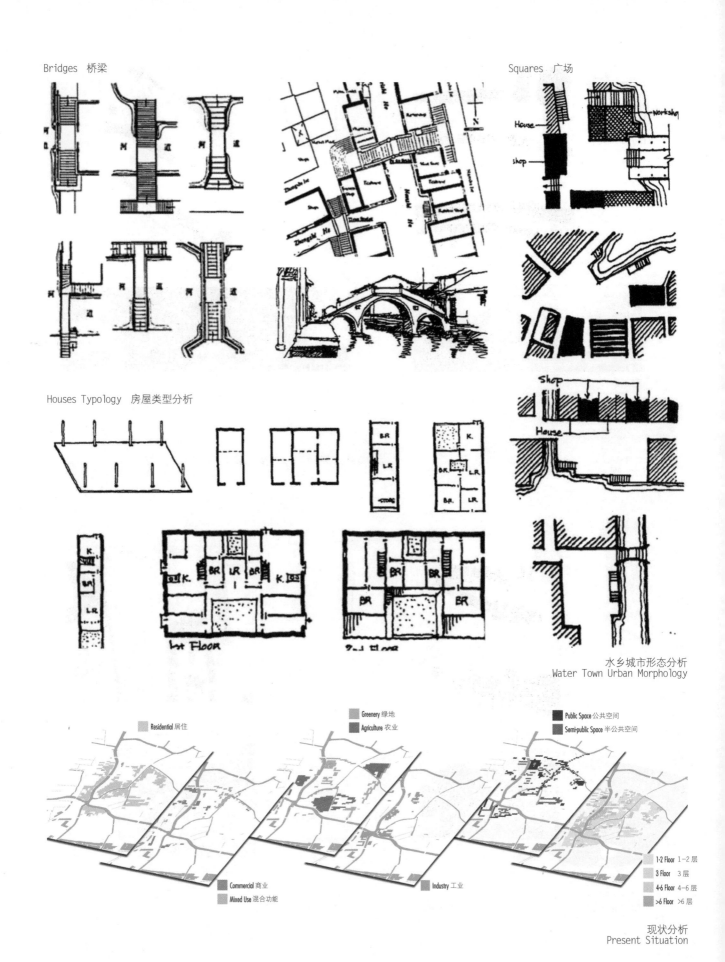

Waterways 河道

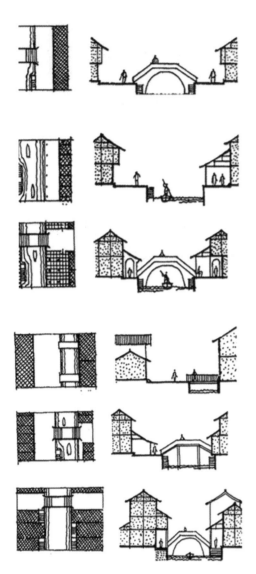

Streets 街道

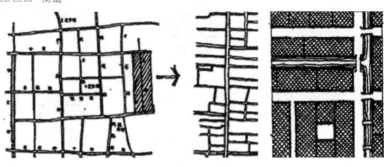

Types of Streets 街道类型

Quays 码头

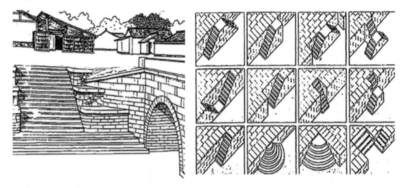

水乡空间要素分析
Space Elements of the Water Town

桥梁
Bridge

A node
Connection to Quays
Mixed use/ Commercial
节点
联系码头
混合功能 / 商业

仓库
Warehouses

Post-industrial Warehouses with the huge indoor public spaces they offer form an opportunity to be reused as community oriented services.
拥有巨大室内公共空间的废旧仓库可改造成为社区服务中心

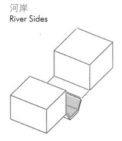

河岸
River Sides

Town backbone
Public space
Activity hub
古镇主干网
公共空间
活力中心

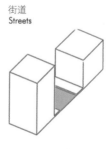

街道
Streets

Main streets: Mixed use/ Commercial
Secondary streets: residential
主要道路：混合功能 / 商业
次要道路：居住

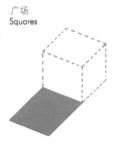

广场
Squares

Small sized around 100 sqm
Usually related to water
Node
Gathering point
100 平方米的小型广场
多布置于水系节点处
聚会节点

方案概念分析图
Concept Analysis

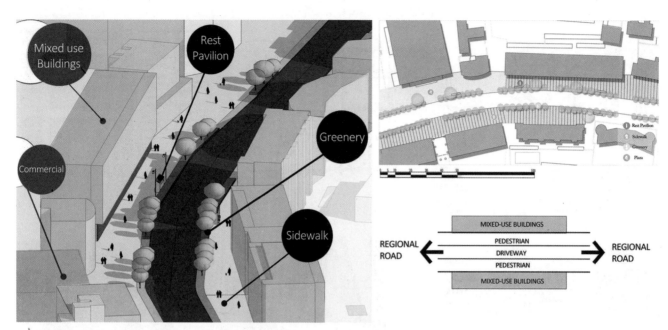

Main Street Guideline 主干道设计导则

Using green belt to seperate cars and people, in order to create a good traffic condition and walking environment. Choosing local plants and the street furniture with local culture characteristics.

运用绿化带隔离人车，创造良好的交通环境与步行环境，选用本地植物和具有当地文化特色的街道设施。

Secondary Street Guideline
次干道设计导则

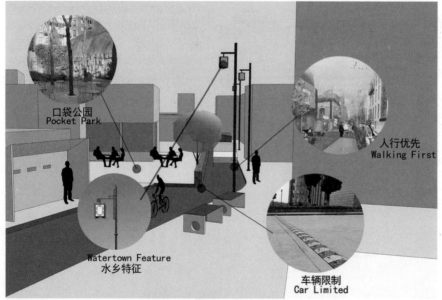

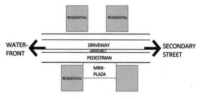

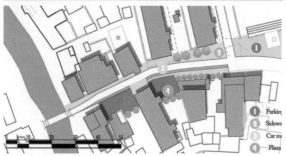

This design aims to create walking priority environment, community service business should be extended into "pocket park" to make the community more cohesion. Meanwhile, the local traditional elements should be used as much as possible.

创造步行优先的社区环境，将社区商业扩展为"口袋公园"，使社区更具凝聚力，同时运用当地的传统元素。

Traditional
Commercial Street
Guideline
传统商业街设计导则

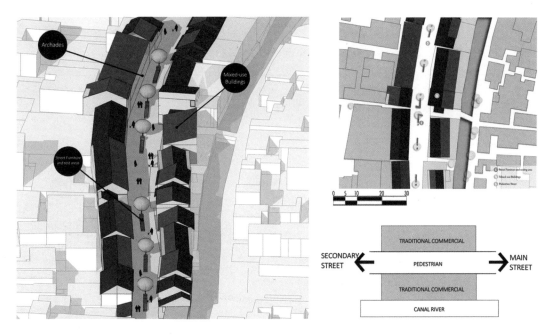

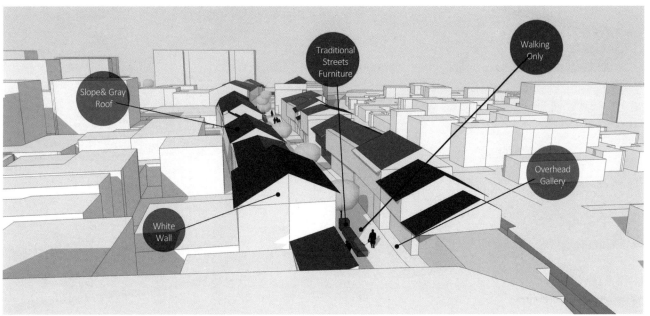

Making use of the traditional texture and traditional buildings of Old East Street, to transform it into a pedestrian street with strong traditional characteristics. Preserving the traditional culture, at the same time, meet the needs of modern life.

利用老东街的传统街道肌理与传统水乡建筑，将其改造成为具有强烈传统特征的水乡步行街，保存传统文化的同时满足现代市民的需求。

 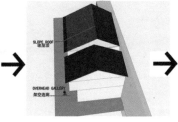   

Community Center & Water Town Creative Industry 社区中心与水乡创意产业园

Turn closed waterfront plant into an open community center and creative industry park with the characteristics of water town, keep the publicity of the waterfront, remain the existing architecture framework as possible, and renovate the facade.

将封闭的滨水工厂转变为社区活动中心与具有水乡特征的创意产业园。保持滨水岸线的公共性，尽量保留现有建筑结构，进行立面改造，使之成为具有水乡特色的社区中心与创意产业园。

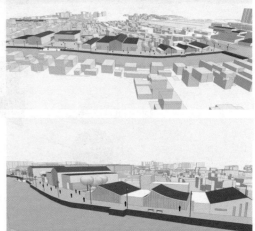

Waterfront Guideline
滨水公共空间设计导则

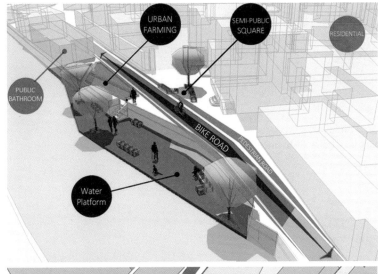

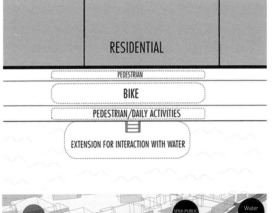

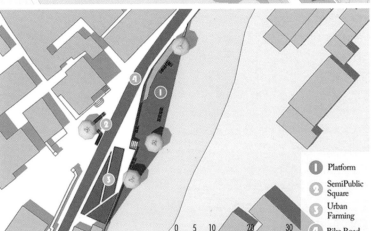

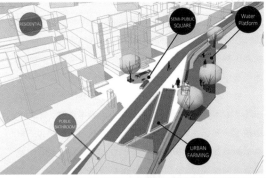

The semi-private character of waterfront area provides a kind of outdoor living space where the residents can sit, eat and chat with their neighbors. Our proposal tries to strengthen this character and develop the quality of the interaction between the people and water.

滨水区域的半私密性导致这里成为一种户外起居空间，居民们可以和邻居一起在此闲坐、吃饭和聊天。我们的方案试图强化这一特征，进而促进人与人之间、人与水之间关系的进一步融洽。

4.3.4

TINGREEN
植绿亭林

学生：Federica Bozzarelli, Alessia Gianfiori, Ana Martin Yuste, Elena Ruberto, Jiangchang Zou

亭林镇正处在快速城镇化进程中，然而这里还保留着一些传统的乡村生活习惯，我们意欲保留并强化这些特征，使得城市兼具乡村的优势。

亭林的外来人口与本地人口具有截然不同的生活习惯。通过实地访谈，我们发现镇中心缺乏一个为全镇服务的社会交往核心空间。由此，我们希望在尽可能尊重现状的情况下，营造一个低密度的镇中心，并在镇中心以城市农园为工具，提供一个社会交往的平台。城市农园同时可以强化亭林的自维持能力。

我们提出了两个传播"自维持的亭林"这一理念的策略：一是在育人农园组织老年人教小孩农业知识；二是沿河开辟一条连接所有节点的科教廊，在此布置废物艺术品、生态标语、环保小知识等。科教廊同时也是激活河岸活力、丰富河岸景观、体现亭林水乡特色的重要策略。

Tinglin is a town that is in the process of being more and more urbanized. However, some rural characteristics are preserved, and our aim is to maintain these and strengthen them even though Tinglin is getting more and more urbanized.

The population of Tinglin consists of local inhabitants and migrant workers whose daily routines differ a lot from each other. By interviewing inhabitants we find out that there is missing a central point for community interaction in Tinglin town center for whole Tinglin. This situation outlines our second aim: we try to create a low density town center, preserving most of the current urban environment, offering a platform for social interaction – using the concept of urban orchards as a tool. By doing this we will strengthen the self-sustainability of Tinglin.

In order to assure a sustainable future for Tingreen, we create an orchard for mixed generation educational project, as well as an educational path along the riverside leading through Tinglin – combining all of our interventions. This includes a reanimation and extension of the riverside, vitalizing the most characteristic element of the water town of Tinglin: the water system.

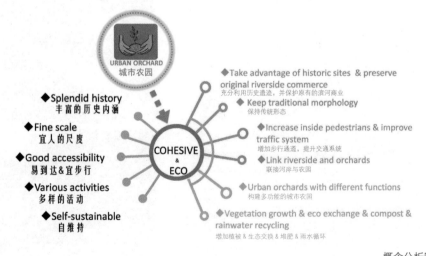

概念分析图
Concept Analysis

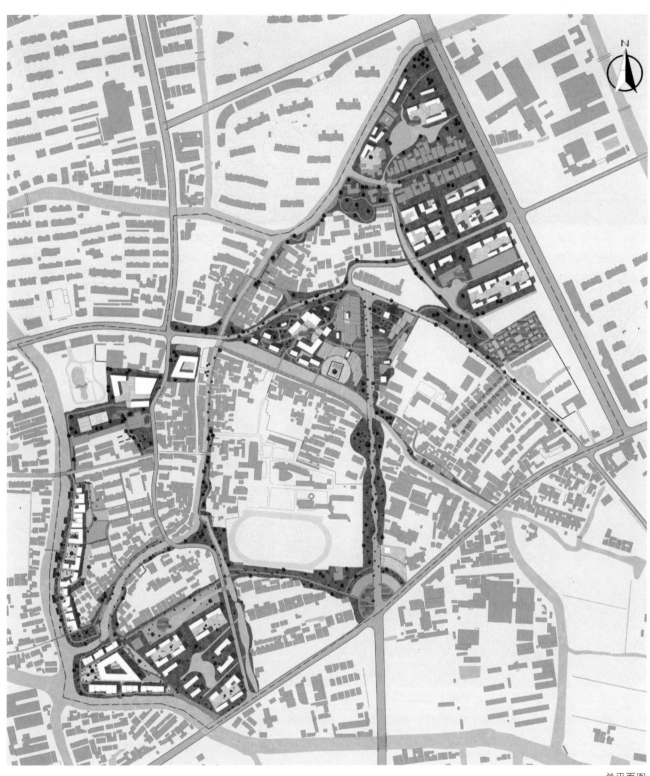

总平面图
Master Plan

Water and Riverfront 水系与河岸开放性与生态性分析

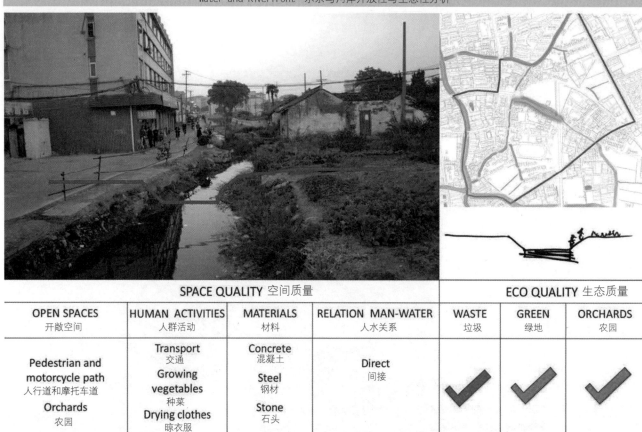

SPACE QUALITY 空间质量				ECO QUALITY 生态质量		
OPEN SPACES 开敞空间	HUMAN ACTIVITIES 人群活动	MATERIALS 材料	RELATION MAN-WATER 人水关系	WASTE 垃圾	GREEN 绿地	ORCHARDS 农园
Pedestrian and motorcycle path 人行道和摩托车道 Orchards 农园	Transport 交通 Growing vegetables 种菜 Drying clothes 晾衣服	Concrete 混凝土 Steel 钢材 Stone 石头	Direct 间接	✓	✓	✓

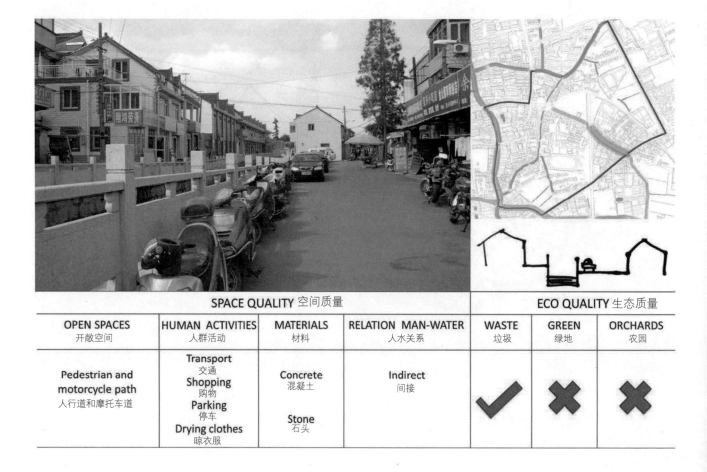

SPACE QUALITY 空间质量				ECO QUALITY 生态质量		
OPEN SPACES 开敞空间	HUMAN ACTIVITIES 人群活动	MATERIALS 材料	RELATION MAN-WATER 人水关系	WASTE 垃圾	GREEN 绿地	ORCHARDS 农园
Pedestrian and motorcycle path 人行道和摩托车道	Transport 交通 Shopping 购物 Parking 停车 Drying clothes 晾衣服	Concrete 混凝土 Stone 石头	Indirect 间接	✓	✗	✗

134

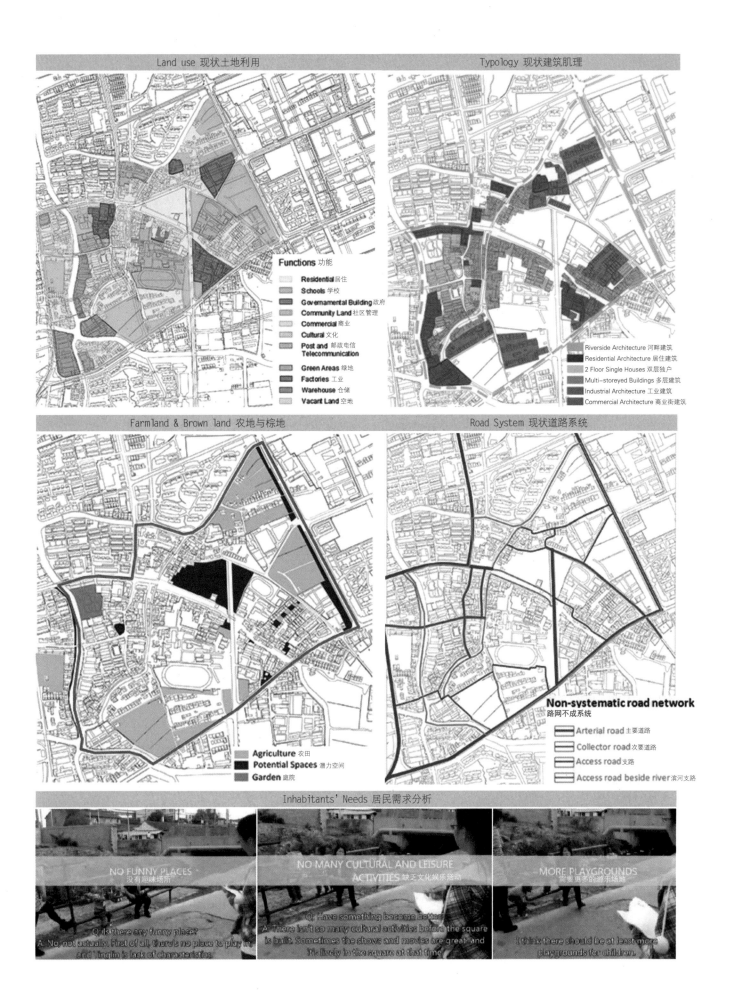

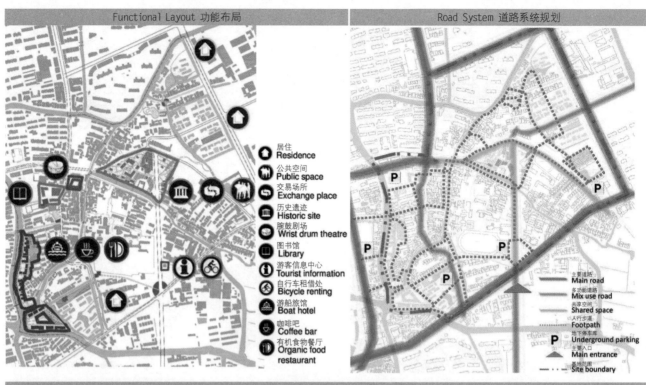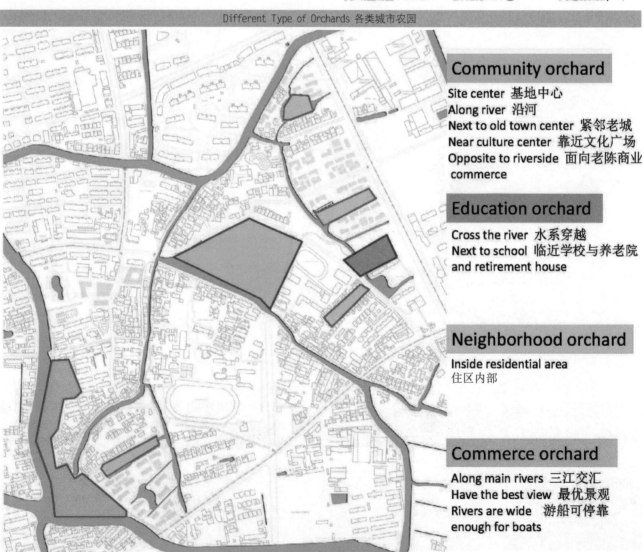

Existing and Designed Situations of Educational Path 科教廊改造前后对比

A Day of Tourists 游客的一天

7:30	Renting a bicycle at main entrance	主入口租自行车
8:00	Visiting community orchard/ aged pine garden	中心农园&古松园参观
10:00	Visiting reading hill	读书堆游览
11:00	Having in organic food restaurant of commerce orchard	赴商业农园的有机餐厅享受美食
13:00	Planting in commerce orchard	商业农园种植体验
16:00	Watching wrist drum performance	赏腰鼓
17:30	Having supper	晚餐
19:00	Check-in in boat hotel	夜宿游船酒店

137

Community Orchards 中心农园

Educational Orchards 教育农园

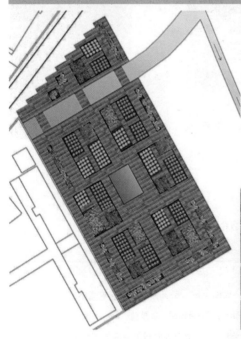

Eco-commerce Orchards 生态商业农园

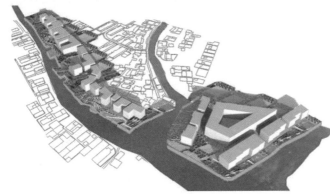

Neighbourhood Orchards 邻里农园

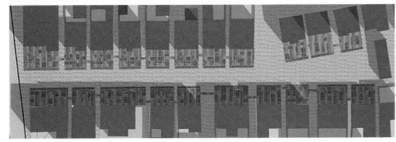

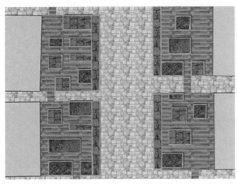

4.4 上海宝山区国际邮轮港地区城市设计
URBAN DESIGN OF INTERNATIONAL CRUISE TERMINAL AREA, BAOSHAN DISTRICT, SHANGHAI

指导老师：田莉，李晴，陈竞姝
INSTRUCTORS: Li Tian, Qing Li, Jingshu Chen

4.4.1 宝山：棱镜城市
BAOSHAN: THE PRISM CITY

4.4.2 河畔之城
ALONG THE RIVER

4.4.3 船"山"
BOAT SHAN

4.4.4 宝山水陆都市
BAOSHAN WATERLAND CITY

宝山基地区位
Introduction to BaoShan Site

吴淞口国际邮轮码头位于上海市北部远郊，宝山新城东侧，是沿江沿海发展带的重要节点。项目基地即为该码头及其邻近区域，用地面积为55公顷。本项目旨在结合邮轮码头的发展，激发基地的活力。

基地距上海中心区约40分钟车程，至虹桥机场约50分钟车程，浦东机场约60分钟车程。基地最近的公共交通站点为地铁3号线宝杨路站，但仍有一定距离，需要约10分钟车程（乘坐公交车需20分钟）才能到达邮轮码头。

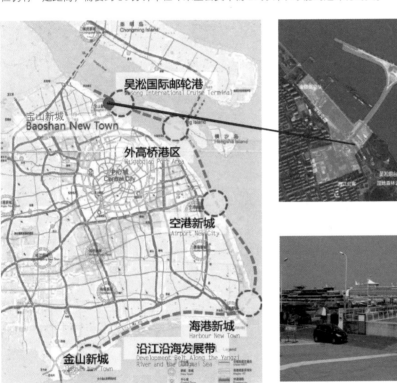

区位二
Location 2

区位一
Location 1

现状照片
Picture of the site

The Wusong International Cruise Terminal is located in the east of Baoshan new town in the northern outer suburban Shanghai. It is a key node along the development belt of the Yangtze and the Huangpu river. The site covers the land area of 55 ha, and includes the terminal and its extended area. The aim of this project is to revitalize the area through taking advantage of the cruise terminal.

It takes approximately 40 minutes from the site to the city center, 50 minutes to Hongqiao Airport and 60 minutes to Pudong Airport by driving. The nearest public transportation is metro Line 3, Baoyang Road Station. However, it takes about 20 minutes from the terminal to the metro station by bus and 10 minutes by driving.

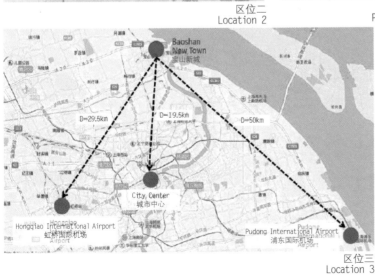

区位三
Location 3

区位四
Location 4

4.4.1

BAOSHAN: THE PRISM CITY
宝山：棱镜城市

学生： Alev Kara, Anikesh Ashwin, Fatma Teber, Ksenia Nikolaeva, Shuting Goh

棱镜城市：以人为本。

随着吴淞口国际邮轮码头的发展潜力不断上升，有必要振兴项目基地，以适应来自各种不同背景的游客的大量涌入，同时兼顾当地居民的生活需求。规划界定三类目标人群，即：国际游客、当地游客和当地居民，采用以人为本的思路，在理解目标人群各自需求的基础上，决定基地的用地功能和城市设计。

形象地说，基地与邮轮码头之间仿佛棱镜的关系，棱镜将光折射出彩虹般绚丽的色彩，不同的色彩折射出各类人群不同的功能需求。项目旨在恢复宝山的活力，使之成为活跃的旅游目的地，通过邮轮产业的辐射，吸引各界人士的到来。

The Prism City: People-oriented Approach.

What is unique about this site is the International Cruise Terminal, locating right beside the site. Due to the rising development potential of this cruise terminal, it is necessary to revitalize this area to accommodate the surge in number of visitors from different backgrounds, yet also not forgetting about the current residents. Three groups of target users are identified and they are the international tourists, local tourists and local residents, respectively. A people-oriented approach is hence adopted to understand what the demands of the users from their perspectives and thereby to determine the land use functions and the urban design of the site.

Graphically, the relation between the site and the cruise terminal forms like a shape of prism, and a prism can be used to disperse light into individual colours of a rainbow. The different colour represents the different functions required by the various groups of people with different needs. It is aimed to make Baoshan come back to life as a vibrant destination through the reflection of the cruise industry.

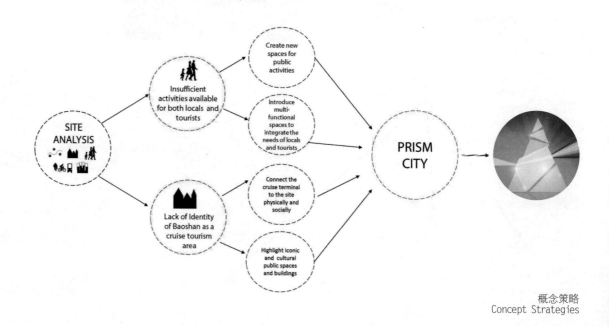

概念策略
Concept Strategies

总平面图
Master Plan

方案概念图　Concept　　方案概念分析　Concept Analysis

详细平面　Detailed Plan

规划目标：

统筹邮轮码头与宝山现有住区的发展；

使宝山成为全天候的活力旅游目的地，吸引各类人群的光顾；

提供丰富公共设施，提升居民生活品质。

滨水步道：

滨水步道是"棱镜城市"这一概念的关键特征，其面向长江和邮轮码头，广阔的开敞空间可以容纳各种不同的功能，并形成人们共同的聚会场所，从而成为吸引三类目标人群的活力地区。

Planning Objectives

To integrate the cruise terminal development with existing Baoshan residential area.

To activate Baoshan as a 24/7 vibrant destination for people.

To improve the urban quality of lives of the people with the provision of mixed amenities.

The Promenade

The promenade along the Yangtze river and cruise terminal would be a vibrant activity space targeting at all the three main group of users. The wide open space provides an opportunity for different functions to integrate together and hence become a common meeting place for all to gather. This promenade is the key feature of our concept of Baoshan as the prism city.

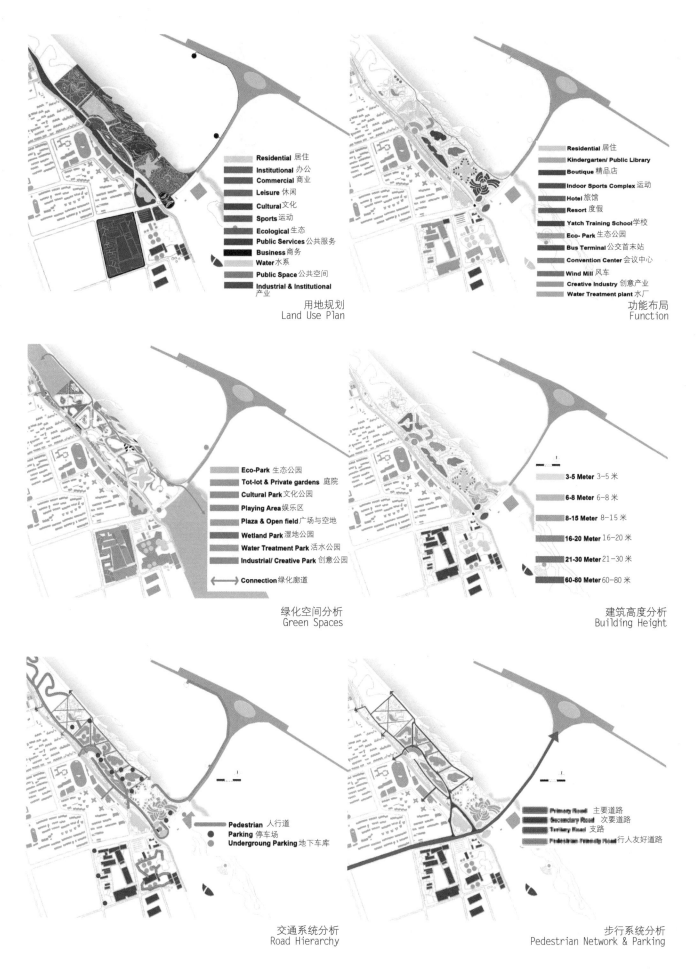

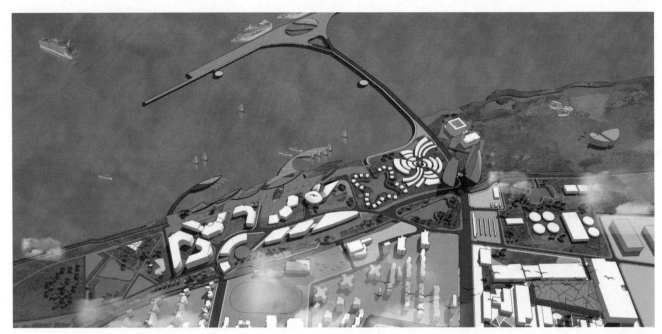

总体鸟瞰一
Bird's-eye View 1

总体鸟瞰二
Bird's-eye View 2

滨水步道剖面
Section of Waterfront Promenade

滨水步道意向
Waterfront Promenade

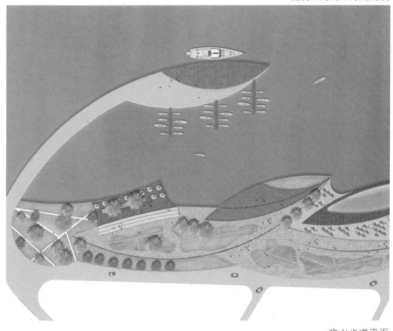

文化长廊意向
Cultural Corridor

滨水步道平面
Plan of Waterfront Promenade

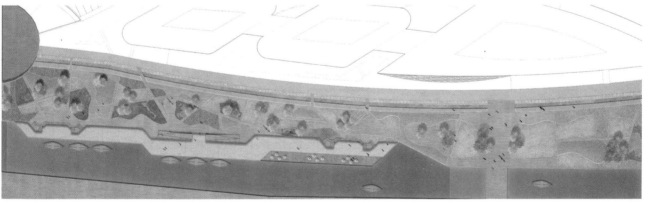

文化长廊平面
Plan of Cultural Corridor

4.4.2

ALONG THE RIVER
河畔之城

学生：Joshua Raff, Vaishali Satyamurthy, Ana Sofia Jimenez, Crantz Simone

本项目基地为上海宝山区一处旧工业码头，规划旨在提出一个以文化活动为主题的空间解决方案。该码头的一个关键特征为它是吴淞邮轮港的终点站，但与当地居民联系不便，居民可从旅游业中获得的机会非常有限。本方案期望通过搭建一个拥有多样文化活动的平台，从而建立一个既能吸引游客，又能为当地居民提供更多机会的新载体。

This proposal is for a former industrial waterfront site in Baoshan district, Shanghai, offers a spatial solution organized around cultural activity. A key feature of the site is the international cruise terminal. Yet the terminal is an entity separated from the local neighborhood. The opportunities for the community to benefit from tourism are limited.

By providing a platform for multiple forms of cultural activity within the site, this proposal hopes to establish a new identity for the area that both attracts tourists and expands possibilities for local people.

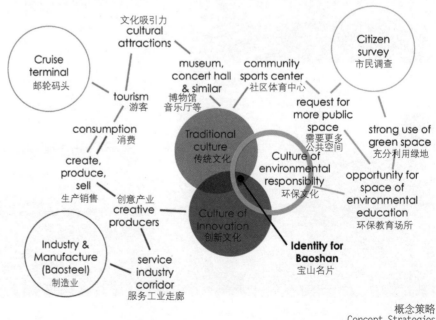

概念策略
Concept Strategies

现场调查
Field Observations

总平面图
Master Plan

Fireworks 烟火晚会

Dragon boat races 龙舟比赛

Sail boating 帆船活动

设计概念
Design Concept

STEP 1: Site re-organization 基地要素重组

STEP 2: Structure the site 构建基地结构

Imperial Palace 皇宫　　　　　The Literati 文人住宅　　　　　Gate & Market 城门与集市

The Qingming Sequence 清明上河图序列

- 现代中国并不需要新的皇宫，那么有什么恰可以与之对应呢？
我们的建议是 **"盛会"**。

–In modern China a new imperial palace is not needed. What then is the appropriate response?
We propose the **EVENT**.

文化设施
– 包括音乐厅、社区中心和传统文化保护机构等。

CULTURAL INSTITUTION

–These are the concert halls, community centers and institutions that maintain the traditional culture.

创意产业
– 包括创意人士、创新文化的先锋等。

CREATIVE PRODUCERS

–In modern Shanghai these are the innovators, the vanguard of a culture of innovation.

The Proposed Sequence 规划序列

STEP 3: Create the event 引入"盛会"

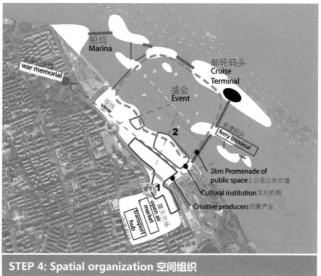

STEP 4: Spatial organization 空间组织

概念生成
Concept Generation

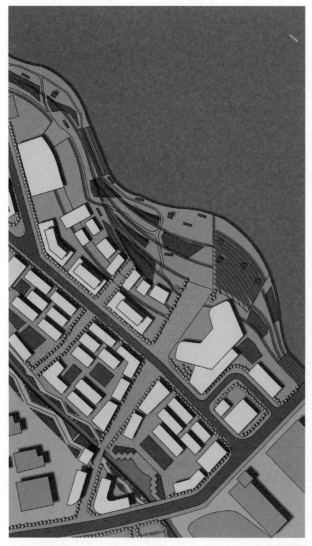

文化区平面
Plan of Cultural District

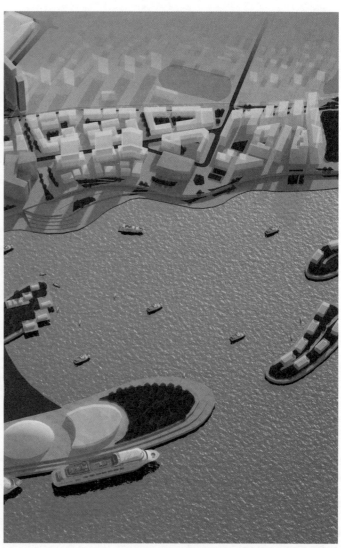

文化区鸟瞰
Bird's-eye View of Cultural District

滨水音乐厅意向
Concert Hall at Waterfront

总体鸟瞰
Bird's-eye View

文化区剖面
Section

门户意向
Gateway

门户平面
Plan of Gateway

4.4.3

BOAT SHAN
船"山"

学生：Daniel Hädrich, Guillaume Hansmann, Vasco Kantowski, Grace Panglu, Alexandre Puech

从国际邮轮码头进入一个城市，意味着为游客们呈现一种欢迎的姿态，这里有大量的购物选择和多种游览项目，但在宝山"吴淞口国际邮轮码头"，情况并非如此。很遗憾，这些优势在该码头及其周边区域严重缺乏。

改变这种状况是本设计的目标。本方案的基本思路为：将邮轮的特色和优势，如紧靠大海的运动场所、娱乐设施和公寓等，引入基地，同时将这些不同的功能分成三组。"大空间"，如电影院和图书馆等，单独设置；而较小且日常的功能，如餐厅、酒吧和商店等，则在每一组内都设置多处。这将形成三个迷你城市，每一个都如同邮轮般建构和运转。

方案最终将包括提升基地现状的策略，以形成一个全新的、更有吸引力的、可持续发展的区域，同时也包括对上海所缺少的邮轮码头建设项目用地的测算以及解决长江沿岸防洪问题的方法。

Entering a city through an international cruise terminal normally means, that you are welcomed by a big gesture, an oversupply of shopping possibilities and many tourist attractions, but not so in Bao-shan at the "Wusong International Cruise Terminal". Unfortunately this terminal and its surrounding area does not have any of this.

To change this situation is the goal of this design studio. The basic idea is to bring the benefits and advantages from cruise ship, like a wide sports and entertainment offer and apartments close to the sea, to the area. Therefore it is necessary to divide the different usages into three groups. The "big functions", like the cinema and library, should be unique in the side, while the small and common usages, like restaurants, bars or shops, are being repeated in every group. The result are three micro cities, which are all working and being constructed like a small cruise ship.

At the end, there should be a proposal, that can improve its situation and create a new, more attractive and sustainable area. Meanwhile we calculate the shortage of available construction land in Shanghai and the issue of flood protection along the Yangtze-River.

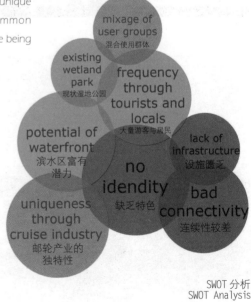

SWOT 分析
SWOT Analysis

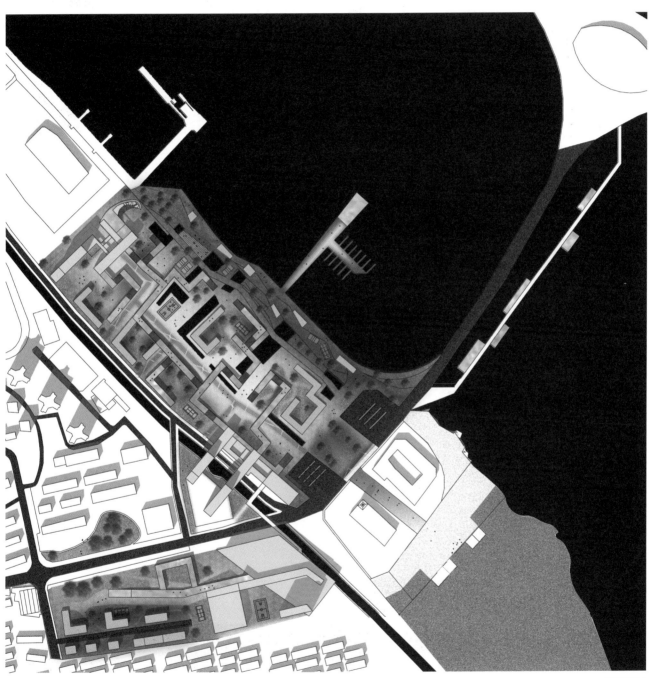

总平面图
Master Plan

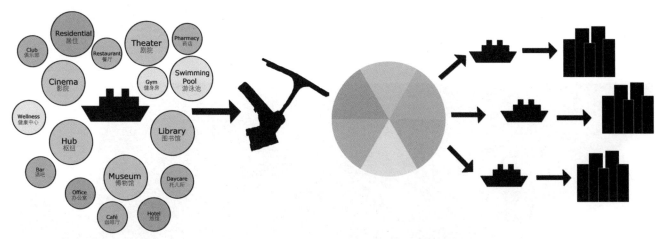

将邮轮的优点和功能延伸至项目基地
Transmit the advantages and functions from a cruise ship to the project side

将功能分配至三艘小船（迷你城市）
Split functions onto three smaller ships (micro-cities)

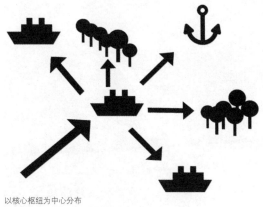

以核心枢纽为中心分布
Distribution from the central hub

拼接不同的使用人群
Mixed user groups

方案概念分析
Concept Analysis

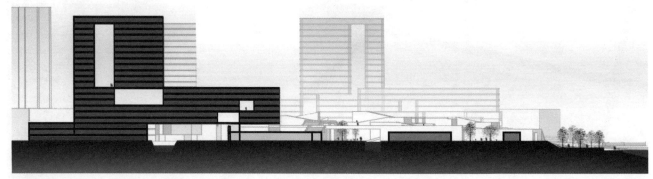

"船体"剖面
Section of Boat

156

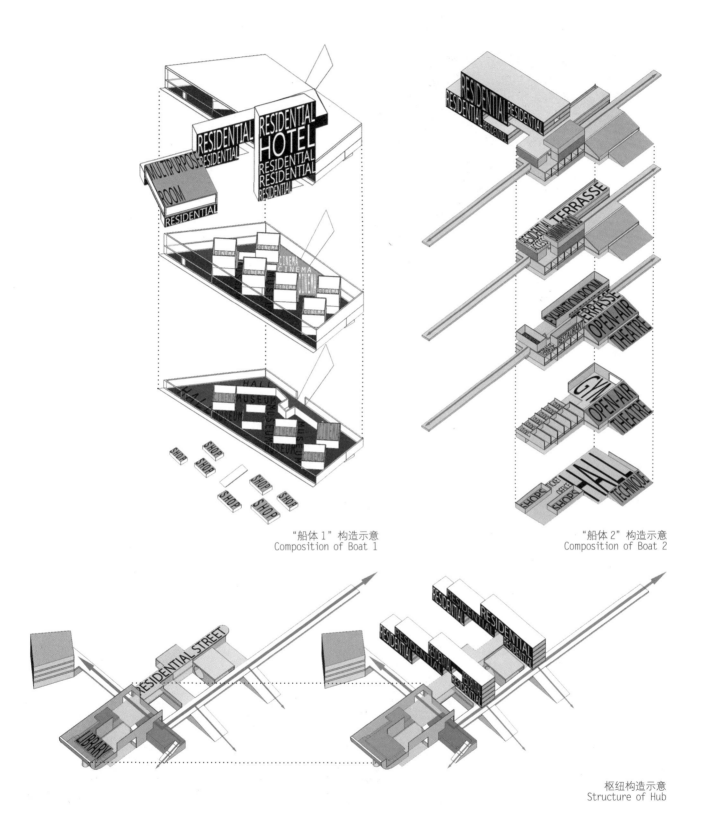

"船体1"构造示意
Composition of Boat 1

"船体2"构造示意
Composition of Boat 2

枢纽构造示意
Structure of Hub

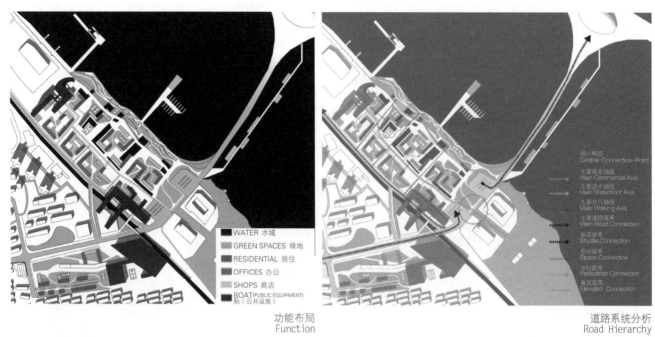

功能布局
Function

道路系统分析
Road Hierarchy

总体鸟瞰
Bird's-eye View

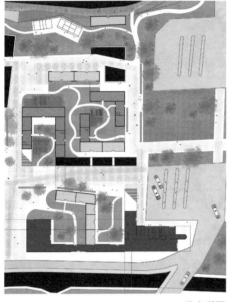

节点详图
Detailed Plan

庭院内景意向
Inside View Courtyard

空中步道意向
View from the Elevated Walkway

4.4.4

BAOSHAN WATERLAND CITY
宝山水陆都市

学生：Carmelo Ignaccolo, Pierre-Baptiste Tartas, Maija Gulin, Veronica Gambaccini, Abdeladim El Gazi, Simone Heath

宝山可谓上海的"北大门"，而方案基地正坐落在长江与黄浦江的交汇处，在上海30公里重点发展区中处于关键位置。该地区的文化和工业/经济特征主要源自抗日战争纪念馆和建于1978年的宝钢。

传统的设计策略是分离式空间设计，以线性方式组织水系统（蓝网）、公园系统（绿网）和道路系统（灰网）。本方案通过更加有机的架构，使空间的使用交叉重叠，将该区域设计成一个更加多元化，更吸引游客驻足并在此享受的区域。相信一旦将这一概念用在宝山，定将使之成为一个令人兴奋的新兴旅游目的地。

Baoshan can be called Shanghai's "Northern Gate" as the site sits at the junction of the Yangtze and Huangpu Rivers. It holds a key position being within 30 km of all key areas in Shanghai. The area's cultural and industrial/economic significances are derived from the Anti-Japanese War Memorial and Bao-steel, founded during the 1978 Reforms.

Traditional design strategies design spaces separately, organize the water systems (blue fabric), park systems (green fabric), and roadway systems (grey fabric) in linear patterns. With a more organic structure, use of space begins to overlap, making this area a more diverse and interesting place to stay and enjoy. It is believed that this concept, when applied to Baoshan, will create a fun and exciting new destination point.

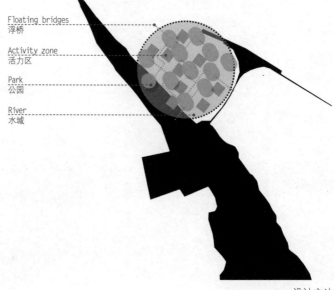

设计方法
Design Approach

总平面图
Master Plan

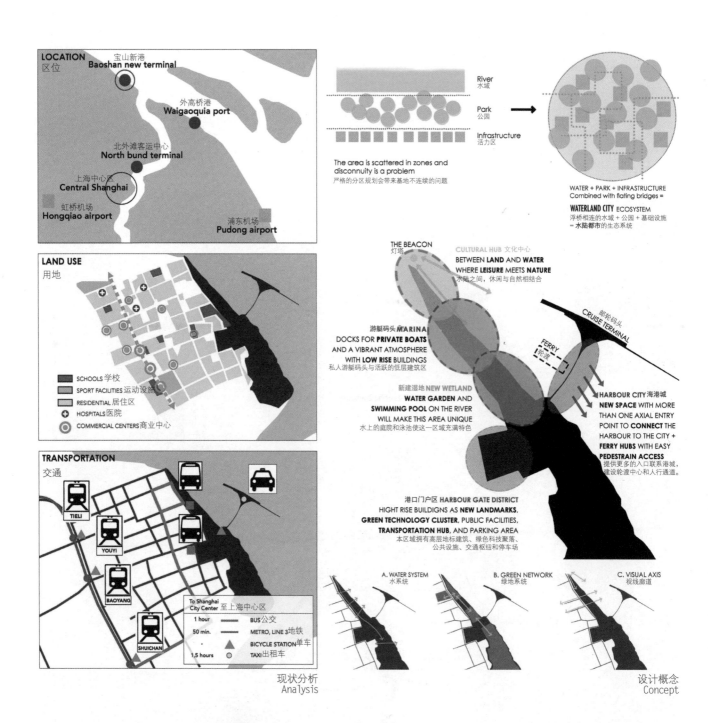
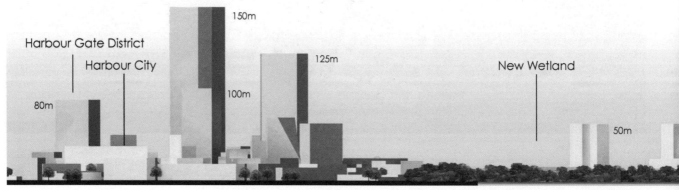

总体鸟瞰一
Bird's-eye View 1

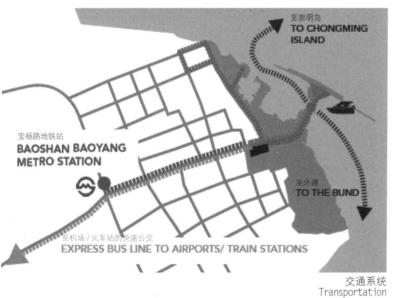

交通系统
Transportation

水－陆生态系统
The Water-land Ecosystem

滨水界面
Waterfront Interface

总体鸟瞰二
Bird's-eye View 2

- Cultural 文化
- Commersial 商业
- Transportation 交通
- Business 金融
- Hotels 旅馆
- Services 服务
- Residential 居住
- Mixed use 混合用地

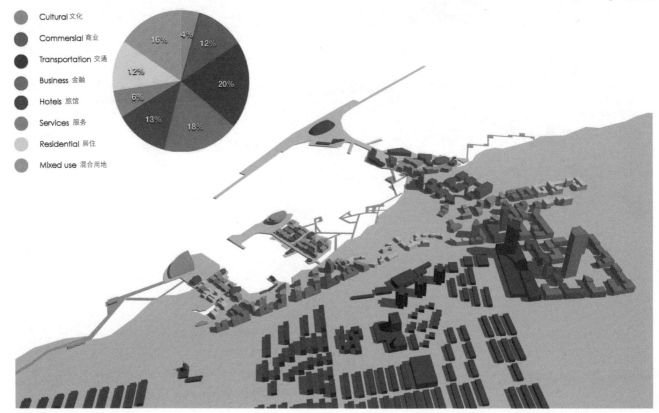

用地规划
Land Use Plan

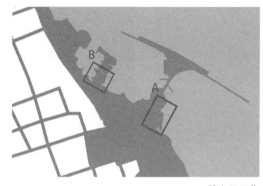

核心区区位
Locations of Key Areas A and B

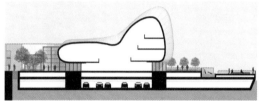

海港城剖面
Section of the Harbour City

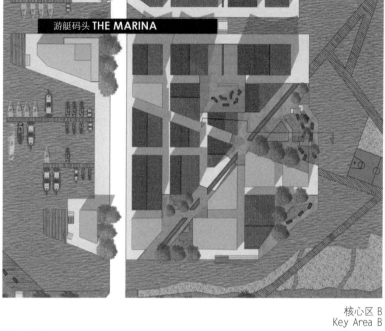

核心区 B
Key Area B

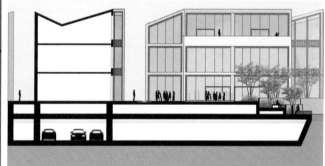

游艇码头剖面
Section of the Marina

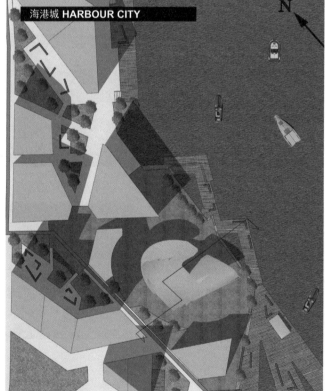

核心区 A
Key Area A

游艇码头意向
The Marina Image

165

Authors
作者简介

田莉，Li Tian

田莉，同济大学城市规划学士和硕士，麻省理工学院城市规划与研究系研究生、SPURS研究员，2006年获得剑桥大学土地经济系博士。现任同济大学城市规划系教授、博导。研究方向为城镇化与土地利用，城市规划与设计等。

Li Tian, Bachelor and Master in City Planning, Tongji University; SPURs fellow of Department of Urban Studies and Planning at MIT; PhD in Land Economy, University of Cambridge. Currently she is a full professor at Department of Urban Planning, Tongji University. Her research focuses on urbanization and land use, urban planning and design.

李晴，Qing Li

李晴，同济大学城市规划硕士，同济大学建筑学博士，2008年芝加哥伊利诺州立大学访问学者，现任同济大学城市规划系副教授、硕士生导师。主要研究方向为社区规划和城市设计。

Qing Li, Master in City Planning, Tongji University; PhD in Architecture, Tongji University; visiting scholar at uic, currently he is an associate professor at Department of Urban Planning, Tongji University. His research focuses on community planning and urban design.

Richard Dagenhart

美国阿堪萨斯大学人类学和建筑学学士，宾夕法尼亚大学建筑与城市规划硕士，现任佐治亚理工大学建筑与城市设计荣誉教授。研究方向为城市设计。

Master of Architecture and Master of City Planning, University of Pennsylvania and Bachelor of Arts in Anthropology and Bachelor of Architecture, University of Arkansas. Now he is Emeritus Professor of Architecture and Urban Design, School of Architecture, Georgia Institute of Technology.

Stefan Al

Stefan Al是一名荷兰建筑师，城市设计师，加州伯克利大学城市与区域规划博士。目前担任宾夕法尼亚大学城市设计副教授。研究方向为城市设计，尤其是城市形态的演化。

Stefan Al is a Dutch architect and urban designer. He got his PhD from Department of City and Regional Planning, UC Berkeley. Now he is an associate Professor of Urban Design at the University of Pennsylvania. He has published various books, book chapters, and articles about the evolution of urban form.

陈竞姝，Jingshu Chen

陈竞姝，同济大学城市规划专业毕业，香港大学建筑学院城市设计系硕士。现就职于上海同济城市规划设计研究院，研究方向为城市设计。

Jingshu Chen graduated from Department of Urban Planning at Tongji University, and then got her urban design master degree from the University of Hong Kong. Now she is working in the Shanghai Tongji Urban Planning & Design Institute as a urban designer.